SOUND UNDERWATER IMAGES

A guide to the generation and interpretation of side scan sonar data.

John P. Fish
H. Arnold Carr

American Underwater
Search and Survey, Ltd

Special Limited Edition

Library of Congress Catalog Card Number 90-63112
Manufactured in the United States of America

ISBN 0-936972-14-9

Lower Cape Publishing
Orleans, MA

Sonar images in this text were processed by, and illustrations were produced by,
Tech-Graphics, Woburn, MA USA

This book is dedicated to the memory of
a great mentor and friend:

Dr. Harold E. Edgerton

Acknowledgements

Many people were responsible for the results in this book. Dr. Harold Edgerton's enthusiasm and brilliance was a continual inspiration. The authors would like to thank EG&G Incorporated for its support in publishing this text. William McElroy's comprehension and explanation of sonar phenomena from our earliest work has been very helpful, and Walter Wienzek provided considerable engineering and technical assistance.

The authors express their sincere thanks to John "Chip" Ryther and William Charbonneau, who wrote the section concerning mosaics and supplied all the sonar mosaic examples. Thomas O'Brien and Kenneth Parolski's advice on deep towing applications and computerized mosaic construction was valuable. James Kosalas provided information on advanced sonar systems and potential future developments. The authors would like to thank Frederick Newton, who graciously provided sonar records and contributed to the description of three-dimensional sonar displays and other computerized sonar processors.

Considerable support was also provided by Harry Maxfield, Paul Igo, and John Spruance in the publication of this text. Steve De Furia and Jim Sullivan furnished creative text design for the book and Terry Snyder provided important engineering and sonar design information.

Ric Walker and Walter Lincoln advanced the authors' understanding of search operations using side scan sonar and high-accuracy navigation systems. The authors would also like to thank R. Story Fish for his contributions to the records in this text and Kevin McCarthy for technical support. Richard Limeburner helped the authors gain a better understanding of computerized drift plotting and its importance for search operations.

Larry Murphy's experience in shipwreck mapping contributed significantly to sonar target analysis. John Thayer provided technical and development assistance. Peter Sachs' oceanographic capabilities and operational management has been supportive during the generation of the records in this text. We would like to thank Hank Leaper for his logistical support during search operations, Mr. Mike O'Rourke for his help in downed aircraft search operations and Larry Webster for assistance in aircraft identification. William Quinn's publication assistance and photographic contribution was important to bringing this text to print. The authors would like to thank Sylvia D. Fish and Laurel Moore for their editorial assistance and Sharon Coe for manuscript typing.

Contents

Chapter 1:
Introduction

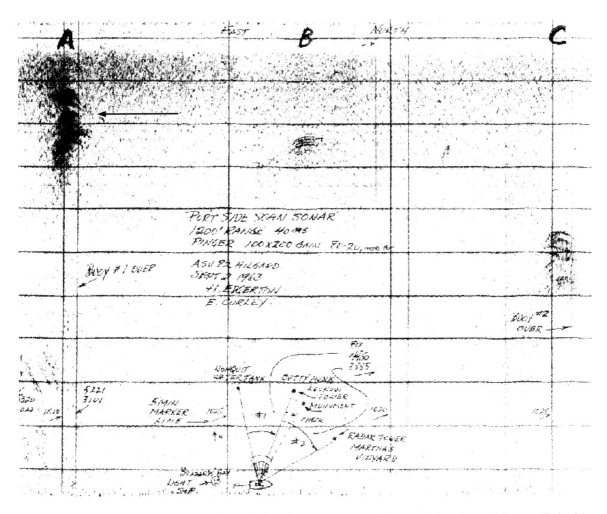

Record 1: Doc's early experiment using side-looking sonar located the mysteriously lost *Vineyard* lightship. The dark hyperbola on the upper left of the record (arrow) is the sonar signal returned from the wreckage. *Record courtesy Dr. Harold Edgerton*

In 1963 Dr. Harold "Doc" Edgerton performed a remarkable experiment with his newly designed "Sub Bottom Profiler." Divers approached him with a request for assistance in finding a lost shipwreck. The wreck held no treasure. Doc had been on many treasure hunts using his sub bottom sonar. The *Vineyard* lightship wreck had simply disappeared during a hurricane in the fall of 1944, with the loss of all on board. The divers wanted to find it and determine why she mysteriously sank. Doc had ideas about widening the single pass coverage of the sub bottom and, with hopes of experimenting on a real search, he agreed. He mounted the transducer of the sub bottom so it could be aimed out to the side instead of straight down. In this configuration, the conical beam would propagate out at an angle perpendicular to the tow path of the support vessel. When they got out to sea, Doc discovered, not unexpectedly, that images of large targets were not very definitive. They consisted primarily of a large hyperbola when the transducers were towed by a bottom obstruction. However, the uniquely mounted profiler did provide the location of one major anomaly lying on the seabed. That anomaly turned out to be the mysterious ship, and is shown in the record on the previous page.

The divers, elated with the find and Doc's latest success, paid little attention to the survey technique that Doc used. But at Doc's firm, EG&G, acoustics engineers began to refine the concept into one of the most remarkable undersea imaging systems ever designed. Although earlier work in long-range, side-looking sonar had been done in England, this experiment provided the groundwork for the development of the high frequency side scan sonar. The new instrument has revolutionized sonar applications and been useful to shipwreck search techniques and seabed imaging worldwide.

Major engineering advances in equipment used for underwater acoustic imaging were made during the 1970's and 1980's. We now have instrumentation that, when applied properly, gives us near perfect images of what lies below the surface of the water. A key to the application of this technology is the proper control and manipulation of the instrumentation to gain the highest resolution and most accurate images. Side scan sonar, although becoming easier to use, is still a fairly complex instrument. There are many options allowing the integration of ancillary equipment and with this integration comes the need for more training and skill to employ the systems properly, for optimal results. The fact that side scan sonar is routinely used from a boat or ship that is subject to the erratic motion of the sea, further compounds the complexity of operation.

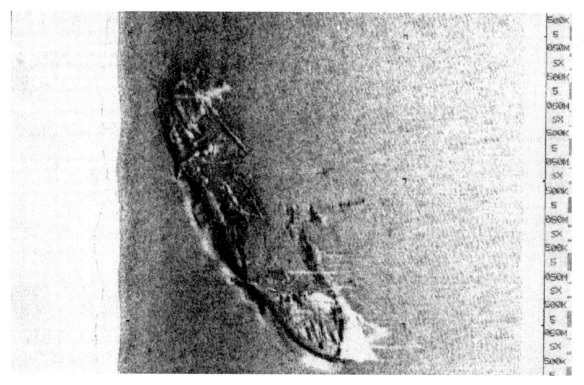

Record 2: Shown above is the slowly deteriorating *Vineyard* lightship, imaged with side scan sonar 27 years after Doc's first experiment. The wreck is surveyed with side scan sonar annually. It shows further loss of structure with each passing year since the detail surveys began in 1979.

The discussion in the following pages focuses on the control of sonar results, rather than detailed examination of how the sonar operates from the standpoint of electromechanical design. In some cases, this text offers the user methods of controlling the sonar system to deal with factors in the environment that cannot be controlled. Furthermore, the text explains unexpected sonar results, and what operators can infer from them.

The majority of sonar phenomena discussed in this text may be experienced with most well designed side scan sonar systems. It is the authors' hope that this text will provide guidance to help sonar operators get more predictable results and be able to accurately interpret unpredicted data.

Chapter 2:
Underwater Sound as a Tool

Illustration 3: Daniel Colladon's work to determine the speed of sound in water during September 1826 is a milestone in the use of underwater sound as a tool for oceanography and sonar imaging. His resulting figures were very close to today's accepted values. *Illustration courtesy* Oceanus

Historically, mariners have realized that sound travels well in the high density underwater environment. In the 1800's, scientists recognized that the speed of sound was consistent and, if predictable, might be a useful tool.

Oceanographers, marine geologists, and archaeologists now depend heavily on sound energy to transform the things we cannot see underwater into numbers, graphs, and pictures, which give us an approximation of what exists in both the shallow murky water of bays, lakes and rivers, and in the deep sea. The instruments that transmit and receive these sound pulses have become sophisticated and very accurate in the past few decades.

Almost all of these systems rely on the accurate prediction of the speed of sound underwater. Underwater acoustics as a research discipline began in 1826, when Daniel Colladon measured the speed of sound in water on Lake Geneva, Switzerland. Colladon positioned two boats 16 kilometers apart. On the first, he fastened a large trumpet fitted with a membrane that would respond to underwater sound. From the second vessel he suspended a bell underwater. On deck was a pan of flash powder and a small flare. The flare was attached to a large hammer also suspended underwater; this assembly was controlled by a bell-ringer in the boat. The procedure of the experiment was to simultaneously set off a bright flash from the powder and hit the bell with the hammer. Colladon's theory assumed that the light from the powder would travel the 16 kilometers instantaneously, while the sound of the ringing bell would take some time to travel the same distance. Colladon, in the second boat, watched for the flash and started a stopwatch when he saw it. He stopped the watch when he heard the sound about ten seconds later.

By today's standards of electronics and high accuracy acoustics, Colladon's methods may seem very crude; however, his empirically calculated value of 1435 meters per second, at a water temperature of 8° C came to within .21 % of the currently accepted value of 1438 meters per second.

In 1899 Arthur Mundy and Elisha Gray were granted a patent for the design of an electrically operated underwater transmitter. Their work led to the formation of the Submarine Signal Company in 1901. The underwater transmitter was refined by this company, and used by the lighthouse service until 1906. In 1912, advances in the transmission of sound underwater led to the first successfully performed echo sounding. However, frequency and beam controlled underwater sound did not really come into wide spread use until high efficiency, coated, piezo-electric "transducers" were developed after 1917. The need to detect and track submarines and surface vessels during wartime led to the development of much more sophisticated sonar. By World War I sound sources were installed on submarines for both echo location and Morse code. In the 1920's and '30's the "submarine oscillator" was used on lightships and other at-sea stations.

The installation of the oscillator on lightships was an interesting application of sonar (SOund Navigation And Ranging). After some rather disastrous collisions between steamers and lightships moored in foggy shipping channels, the oscillator was put into use. This sound generator was attached to the hull of the moored lightship. Simultaneously, the lightships' RDF radio beacon was transmitted, the fog horn was sounded, and the underwater oscillator was activated, emitting an underwater acoustic pulse. Ships approaching the lightship, if fitted with the proper equipment, could receive all three signals. To receive the underwater sound, some ships were fitted with an "ear" mounted in the hull. If the receiving

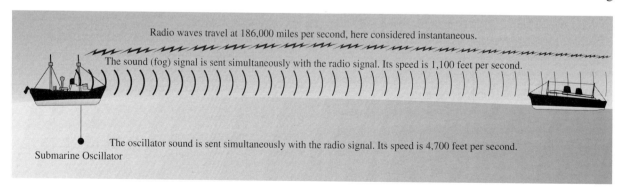

Radio waves travel at 186,000 miles per second, here considered instantaneous.

The sound (fog) signal is sent simultaneously with the radio signal. Its speed is 1,100 feet per second.

The oscillator sound is sent simultaneously with the radio signal. Its speed is 4,700 feet per second.

Submarine Oscillator

Illustration 4: The development of the submarine oscillator gave ocean going vessels a method of determining the range to a lightship thus helping to prevent collisions. Since the speed of sound in air and water were known, the range was determined by timing underwater and above water signals transmitted by the lightship.

hull was fitted with two listening devices, one on each side, stereo sound allowed the listener to determine some directionality. Further, by timing the receipt of a bell, horn or radio beacon transmitted in the air, in relation to the receipt of the signal from the underwater oscillator, it became possible to determine range to the source.

Almost all sonar systems fall into the category of either "passive" or "active". The passive sonar is able to only receive sound and listens to sounds generated externally to itself (such as a submarine or the acoustic output from ocean mammals). It receives these signals through a hydrophone. A passive sonar deals only with sound that has one-way travel time; the time it takes the sound to travel from the source. Active sonar generates a signal, transmits it through the water where it will reflect off targets. These reflected echoes are then sensed by the active sonar's hydrophone and received. Virtually all side scan sonars are active systems.

Illustration 5: The collision between the *Olympic* and the *Nantucket* lightship was typical of the incidents that struck fear in the hearts of lightship crews. Following her course to New York from Europe, the liner cut into the lightship in thick fog. The *Nantucket* sank in forty fathoms with seven of her crew. *Illustration courtesy Paul Morris*

The components that make up an active sonar system, the projector and the hydrophone, are called transducers. In this text we will often refer to these transducers which are commonly mounted in the side scan sonar towfish. The projector transforms electrical signals into underwater pressure waves (sound). The hydrophone performs the opposite function. In side scan sonar, transducers can transmit sound, receive sound or do both. In some applications it is advantageous to have two separate transducers performing the two different transmit and receive tasks, but most medium frequency side scan sonar systems utilize transducers that both transmit and receive sound signals. Shaping the sonar beam is accomplished through sophisticated transducer construction and sonar can be designed to transmit a great variety of beam types and shapes. Transducers can transmit omnidirectional or directional beams and they can emit pulsed or continuous signals. Also power levels can be controlled. As we will see later in this text, transducers used in side scan produce a very specially shaped beam.

Sound propagates in the form of a wave and in order to demonstrate how this works we should examine a wave form that is visible to the eye. Dropping a small, tennis ball into a into a pool of water would cause waves on the water's surface. These waves would propagate outwards, away from the ball. If we alternately push down and release the ball, we can continue to produce these waves as long as the motion is not stopped. Now, if we examine these waves closely we can categorize a number of their features. The number of waves that pass any specific point in one second is known as the frequency of the wave. Frequency is measured in cycles (waves in the above case) per second. One cycle per second is called one Hertz (always capitalized because it is named after H. R. Hertz). The distance between the waves (measured at the same point in each wave) is the wavelength and the rate at which the waves pass determines their speed and in the case of sound waves underwater, we accept approximately 1500 meters per second as the sonar velocity. If we push down on the ball at a faster rate, the frequency increases and the wavelength decreases.

Although the example above is very simplified, the two features of sound waves described, wavelength and frequency, are important in side scan sonar. For underwater sound, only one of these needs to be specified to determine the other. As we will see in later chapters, high-frequency sound energy is greatly reduced by seawater. Low-frequency sound energy is reduced at a far lesser rate. For instance, a sound pulse of 50 Hertz can be transmitted many thousands of kilometers in the ocean, but a pulse of 100 kHz, a common side scan sonar frequency, can be transmitted only 1 or 2 kilometers. In today's active sonar and side scan systems, approximate two-way working ranges are as follows:

Frequency	Wavelength	Distance
100 Hz	15 meters	One thousand kilometers or more
1 kHz	1.5 meters	One hundred kilometers or more
10 kHz	15 centimeters	Ten kilometers
25 kHz	6 centimeters	Three kilometers
50 kHz	3 centimeters	One kilometer
100 kHz	1.5 centimeters	600 meters
500 kHz	3 millimeters	150 meters
1 mHz	1.5 millimeters	50 meters

This relationship between frequency and range has a wide variety of connotations. If the side scan sonar user wants to transmit, and receive, sound pulses at long ranges (and cover wide areas in a short time) a low frequency source is best to use. Unfortunately, the low frequency sound has longer wavelengths and often longer pulse widths (the amount of time the sonar is active). This provides lower resolution in the resulting information. If the need is to accurately image fine details, then it is preferable to use a higher frequency sonar. But these short wavelengths cannot be transmitted long distances and thereby limit the usable range. Often, the sonar operator will find that he must choose the usable frequencies for his application that will provide the best trade off between range and resolution.

Power levels of transmitted and received underwater sound are typically expressed in decibels (abbreviated as dB). These units give the sonar engineer an appropriate logarithmic scale with which to measure sound levels. Sonar output levels are usually noted as they relate to a reference level such as one microbar or micropascal at a specific distance (often one meter). Most well designed side scan sonar systems have adequate output for use in the applications described in this text but, as mentioned, these power levels vary widely. In general, coastal depth systems have an output power level in the region of 225 dB at a reference of one micropascal at one meter. For specific information on any particular sonar system, the user should contact that manufacturer.

Chapter 3:
Applications of Side Scan Sonar

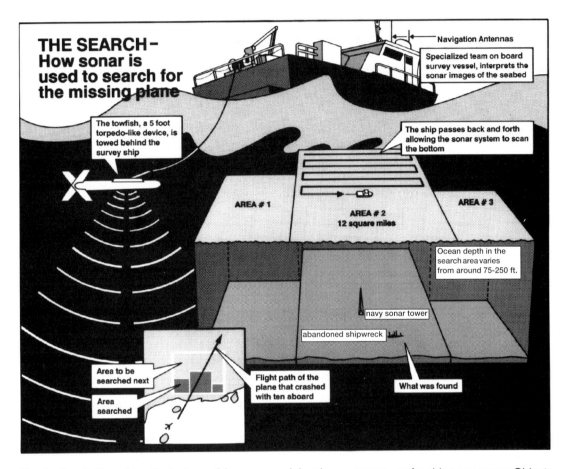

THE SEARCH–
How sonar is
used to search for
the missing plane

Navigation Antennas

Specialized team on board
survey vessel, interprets the
sonar images of the seabed

The towfish, a 5 foot
torpedo-like device, is
towed behind the
survey ship

The ship passes back and forth
allowing the sonar system to scan
the bottom

AREA # 1

AREA # 2
12 square miles

AREA # 3

Ocean depth in the
search area varies
from around 75-250 ft.

navy sonar tower

abandoned shipwreck

Area to be
searched next

Area
searched

Flight path of the
plane that crashed
with ten aboard

What was found

Illustration 6: Searching the bottom of the ocean or lakes is a common use for side scan sonar. Objects from a few centimeters to hundreds of meters in size can be pinpointed and identified with sonar. As shown in this illustration, the sonar is towed along parallel track lines. Each swath of the sonar beam overlaps the last. By this method, large areas can be effectively covered in a short time.

TARGET LOCATION

Side scan sonar will very effectively image large areas of seabed. Because of this, the sonar is commonly used to find specific targets in a large body of water. When a vessel goes down, or an aircraft crashes into the water and sinks, a systematic search with a side scan sonar can pinpoint the target with high accuracy. In addition to using side scan to pinpoint the target's exact location, the sonar may indicate the condition of the craft, how it is being affected by its environment, and its current profile above the seabed.

The side scan is very useful in locating historic shipwrecks and other cultural sites of interest to archaeologists. Often, however, these targets have a small acoustic profile and this would require high side scan frequencies to properly image and detect them. Search operations for these older sites typically require short sonar ranges, thereby reducing the coverage per pass and increasing the amount of time required to effectively cover any given area.

DETERMINATION OF SEABED CONFIGURATION

The ocean floor contains a great variety of structures. Some areas are covered with miles of flat ooze, while others contain precipitous rock outcrops. Flat, desert-like sand deposits occur in some areas, while clay, gravel, or round stones blanket others. Side scan sonar is an ideal tool for mapping these areas for a number of applications. In a standard towing configuration, side scan delineates even the smallest change in the sea bed, whether it is a very slight incline, a small depression or sinkhole, or a change in sediment deposition due to a change in the bottom structure. These surveys are performed in deep as well as shallow waters. Transoceanic cables must be laid on routes that avoid seamounts and other obstructions while taking an economically short path. Wide swath sonar systems are used by international telephone companies to determine these routes.

PETROLEUM INDUSTRY APPLICATIONS

Overall surveys are easily performed with side scan sonar as well. In the 1970's and early 1980's, petroleum interests became more concerned with the seabed environment as more offshore rigs were placed into action, and pipelines that carried the petroleum product from one place to another became more common. In some underwater environments, such as the Gulf of Mexico, certain areas are more stable and otherwise more suitable for offshore rig placement than others. A large scale, slant range corrected sonar map aids in the proper placement of an oil drilling, or production rig.

In some underwater areas, such as those at major river mouths, sediment buildup sometimes results in "slumps" or landslides underwater. It is very important not to place any seabed structure such as an oil rig in these very unstable sediments. Side scan sonar, when properly employed, assists in determining the stability of seabed areas. The placement of pipelines on the seabed requires the same care because pipelines may be in use for many decades. Mapping records created from side scan sonar have assisted in developing advanced methods of laying pipelines. The Gulf Of Mexico was established as a petroleum producing area long before the North Sea in Europe. Pipelines were frequently laid in the Gulf in the 1930's and 40's. These were laid with little concern for obstructions and protrusions on the seabed. Side scan examination of many of the pipelines provided sonar records that have been described as "...a random placement of spaghetti", with pipelines crossing one another many times. Upper ones had been laid in later years with no knowledge of the earlier pipe positions. By contrast, many of the North Sea pipelines were laid more recently when technology in undersea, high resolution imaging and robotics was used. These pipelines have been constructed on supporting bridges wherever they cross those that were laid earlier. Also, pipe bridges supply support where they cross a large ditch, thus preventing "freespan" which could weaken and break a pipeline. Side scan sonar has been very useful in mapping the position of all the pipelines and determining areas where bridges will be needed in a planned pipelay project.

DREDGING

Dredging is another industry where side scan sonar is frequently used. In areas where underwater sediment builds up, or rock outcrops are present, blasting and dredging operations are often required to clear an underwater path, to allow the passage of vessels. Side scan imaging can show waterways authorities what areas require dredging and indicate the type of material that needs removal. Bathymetric surveys are also helpful with this work, and are crucial for exact depth measurements. However, side scan sonar is being used more and more to provide an overall picture of the area requiring excavation for pre-dredging operations.

After dredging operations are complete, side scan sonar is helpful in demonstrating the extent of the work completed. The actual scour of the dredging head left in the sea or river bottom is clearly indicated in the minute topography of the bottom image. As a dredge head sweeps across the sediment being removed it alters the bathymetry, and although a bathymetric survey is usually required, all of this can be surveyed and inspected for coverage with side scan sonar.

MINEHUNTING

Since the mid 1700's when the first mines were placed underwater as a threat to coastal shipping, detecting and pinpointing the location of these mines has been a foremost concern for those responsible for protecting the waterways. The mine has been referred to as the "silent killer" and, when a waterway has been mined, it is considered closed to major shipping if the mines cannot be located and neutralized.

Recent side scan sonar technology has provided very high resolution systems that detect and pinpoint the location of underwater mines. The worldwide practice of "minehunting" has employed side scan sonar as one of its primary tools. The instrument has also played an important part, after an area has been mined, in channel clearance, and "Q-Route" (areas of safe passage) selection. Although ROV's (Remotely Operated Vehicles) are most often used for mine "neutralization" or elimination, side scan is one of the primary detection tools for mines.

Although we often think of mines as the early World War I and II large, black spheres with multiple horn-like protrusions, in truth, mines are manufactured in a variety of sizes, shapes, and for a variety of tasks.

Mine manufacturers are now making mines that are harder to detect, which have sloping surfaces that provide very low reflecting faces, and materials that absorb much of the sound incident to it. To an experienced sonar operator, the latter case, which involves a ghostly shadow in the seabed where sound is absorbed leaving no returning signal to the sonar, would be almost as suspicious as a hard target with high reflectivity.

The side scan sonar is an ideal tool for both channel conditioning and mine target acquisition and is being used around the world by naval forces with an interest in harbor defense.

ENVIRONMENTAL APPLICATIONS

Side scan is also a useful tool in examining the cause and effect of pollution on the marine and fresh water environment. Outfalls, legal and illegal dump sites, and effluent plumes are all frequently inspected using side scan sonar. Since sound waves emitted by the sonar are affected by density variations in the water column, and many of the pollutant plumes are of different densities, sonar often images these very accurately. With a side scan sonar record of these anomalies, the surveyor can determine the source, deposition, and distribution of the effluents.

Environmental effects of thermal pollution became a concern in the 1970's, as power plants became more common along waterways around the world. Inspection of these thermal plumes have been effectively performed by environmental engineering firms using side scan sonar. Since water of a different temperature has a markedly different density than surrounding water, sound waves from sonar are dramatically affected by them. The side scan shows this effect with extremely high acuity and will delineate thermal plumes as they are carried into an existing body of water.

FISHERIES

Fisheries interests have utilized side scan sonar in a number of ways to both increase catch and determine fish population densities in specific areas. Although much of the body of individual fish species has little difference in density from water, today's side scan sonar has the capability to image relatively small (< 25 cm) fish within schools of individuals. The shadow of fish in mid-water is often cast on the seabed at some distance away from the image of the fish and this information can be used to determine the height of the fish above the seafloor.

Artificial reefs is another structure that is effectively surveyed using side scan sonar. Man made reefs will slowly degrade and scatter, becoming less and less effective over time. Reefs located in water less than fifty feet deep are periodically subject to storm surge, or severe currents. This reef disintegration process is monitored using side scan sonar while simultaneously recording fish population density in the region.

Shellfish beds are also located and monitored for size and conditions, using high frequency side scan sonar. An experienced sonar operator can identify shellfish beds by a change in contrast (e.g. a bed of mussels is a dark smudge on the record) when he surveys an area that is suspected to maintain such beds. Additionally, sonar can assist biologists or fishermen in determining if a bed has become over fished.

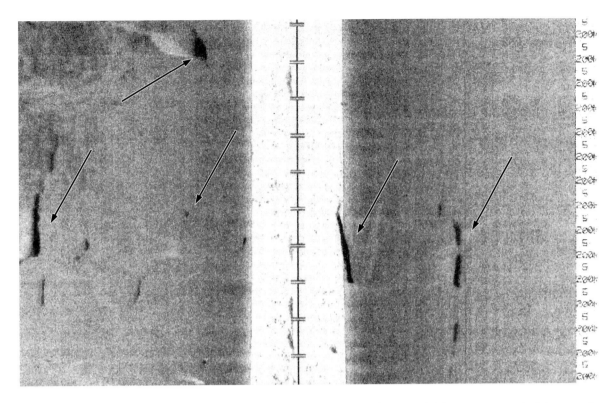

Record 7: Fish schools can be identified and quantified through the use of side scan sonar. In this record, densely packed groups of herring (*Clupea harengus*) are shown swimming in mid water. The position of shadows often help fisheries scientists determine vertical distribution of the schools. Fish are typically not a problem for the surveyor, but when they occur in dense schools, they can mask bottom targets.

Chapter 4:
The Record

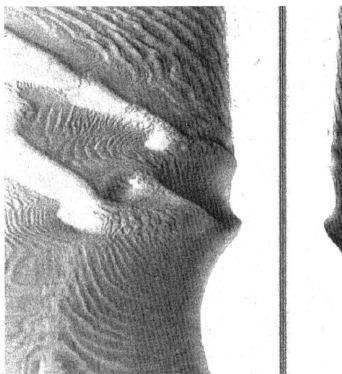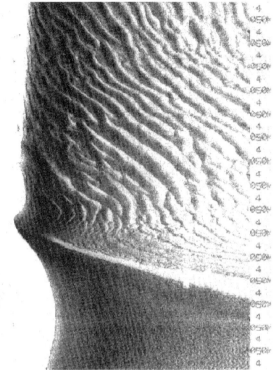

For those readers unfamiliar with side scan sonar records, a brief introduction is provided. This introduction is intended to help the reader understand how the records are made and what parts of the real world environment cause the variety of images and shadows on the records.

The side scan functions in the following manner. The transducer assembly is towed on a steady course and at a constant depth through the water (see Illustration 6). During towing, the assembly emits sound pulses at precise and regulated intervals. The system receives the returning echoes from the water column and seafloor immediately after emitting a pulse. This continues for a short period of time until the next pulse is transmitted, and the cycle begins again. The returning echoes from one pulse are displayed on the recorder as one single line, with dark and light portions of that line representing strong or weak echoes relative to time. Similar to a TV screen, which is made up of hundreds of lines, any single line does not provide the viewer with significant information. However, since this process is repeated hundreds of times per minute, the resulting lines, which are juxtaposed in the sonar display, form a coherent picture for the operator.

Adequate interpretation of side scan sonar data requires some knowledge of sonar phenomenon. There are many variables that affect the sonar data and how the resulting sonar records depict the seabed and various targets. Environmental conditions such as wind, waves, currents, density gradients from temperature or salinity changes, all affect the quality and interpretability of the sonar data.

Operator management of the system and the data gathering process also severely impacts the data. Survey vessel course, tow speed, towfish altitude above the seabed, and range or tuning selections will dictate the quality of the sonar data. This text provides some examples, and explanations, to assist the side scan sonar user to more accurately interpret data.

The display on the recorder (or video unit) is three *channels* of data. Two of these are the port and starboard side scan channels and the third provides the operator with an area for event times, navigational data, and water column display. This third channel is very useful when the system is being operated in the corrected mode. During this time, when the system is generating slant range corrected data, the water column is removed from the port and starboard data channels. The third channel provides a visual display of the towfish height above the bottom. It also displays uncorrected data from either the port or starboard channel. This not only allows for bottom tracking verification, but also displays valuable water-column data such as, suspended sediments, targets, and sea-surface effects.

Record 9 and Illustration 10 describe the anatomy of a side scan sonar record as it is displayed in the uncorrected mode. The sketch represents a vertical *cross section* of the operating environment, while the actual record represents the seabed in a horizontal plane as if viewed *from above*. The analysis of this record is reasonably straightforward, if you remember that side scan sonar technology is based upon the elapsed time between the outgoing pulse and the returned reflection from targets in the environment.

The two lines on the record at **A** are the *trigger pulse*, or the first sign of the outgoing acoustic pulse, on the port and starboard channels. The thin line between these two, in the two dimensional display of sonar, represents the path of the towfish. When this particular record was made, the towfish was towed closer to the surface than the bottom and the wide beam of the sonar struck the surface before the bottom. The printed echo of the surface on the record is called the *first surface return*. In this example record, **B** is the first surface return. The sea surface is a good reflector, and may be visible on the sonar record regardless of whether it is displayed near the outgoing pulses or at some range further out on the record. The first surface return is typically a good indication of the towfish depth and this distance can be scaled from the record using the 25 meter *scale marks* (**J**). The depth of the fish should be scaled from the trigger pulse to the first surface return. In this record the fish was towed approximately 3 meters below the surface. **C** is known as *sea clutter* and is caused by the sea surface reflecting portions of the sonar beam. Sea clutter is more noticeable on the starboard channel than on the port channel. The sea surface is a better acoustic reflector on that side of the towpath. This is often due to the fact that surface waves are better reflectors on the downwind side than they are on the upwind side. **D** is known as the *first bottom return*. Except for very soft bottom conditions such as mud, the bottom is almost always a strong reflector. This first bottom return is a good indication of towfish height. This distance is scaled in the same manner as the towfish depth with the exception that the scaling is done from the trigger pulse to the first bottom return. Here the fish was towed about 7.5 meters above the bottom. In certain applications, the sonar system uses this altitude information for processing the data prior to printing it on the record. **E**, the white area between the outgoing pulses and the

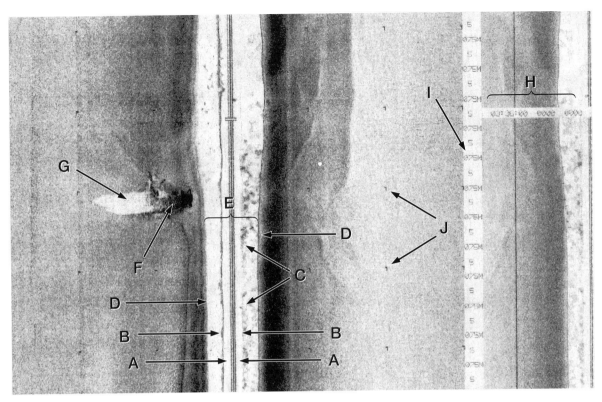

Record 9: The portions of this record are as follows: A: trigger pulse, B: first surface return, C: sea clutter, D: first bottom return, E: water column, F: sunken fishing vessel, G: shadow, H: Data channel, I: System operational settings, J: 25 meters scale marks.

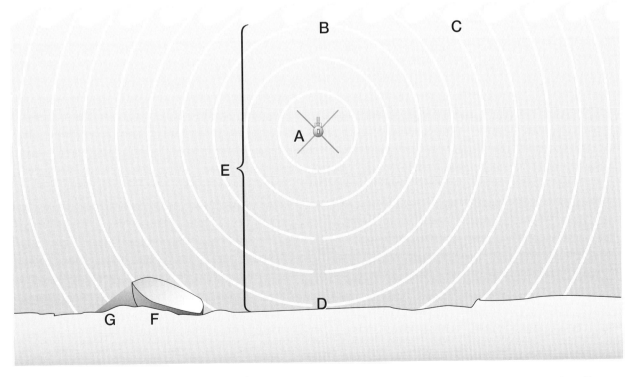

Illustration 10: This sketch describes the conditions under which record 10 was made: A: trigger pulse, B: first surface return, C: sea clutter, D: first bottom return, E: water column, F: sunken fishing vessel, G: shadow.

first bottom return is known as the *water column*. This area of the sonar record is not displayed in some operating modes. **F** is a sonar target lying on the sea bed. In this case it is a shipwreck; however, rocks and other debris will often present good reflectors on the bottom and show up as darker areas on the record. Since almost all of the acoustic energy from the sonar was reflected back from the target there is little *insonification* just behind it at **G**. **G** is the acoustic shadow cast by the target. This area is displayed very light on the record, because almost no acoustic energy from the towfish reached it. Areas **H** and **I** on the sonar record are determined by the recorder, and therefore cannot be shown in the sketch. **H** is the third channel, that shows the towfish altitude when the water column is not displayed on the record. This channel is very useful in tracking the altitude of the fish, as well as providing space for record annotation without overprinting the primary part of the sonar record.

Records in this book will be cropped in a variety of ways depending on the record content. There is also a mix of records with and without the water column in the center of the record (corrected and uncorrected records) in the text. In most cases, the altitude channel was cropped out prior to publication to provide a larger record image for the reader.

Chapter 5:
Theory of Operation

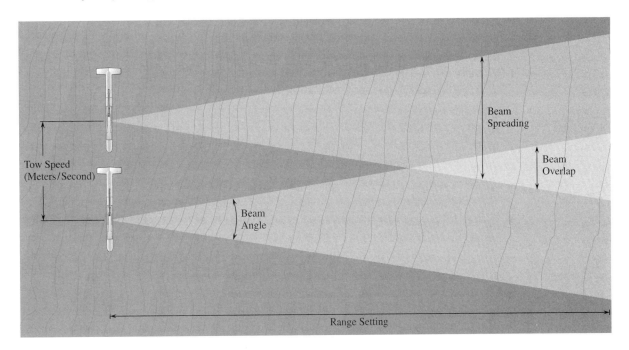

Side scan sonar has been defined as an acoustic imaging device used to provide wide-area, large-scale pictures of the floor of a body of water. Simply stated, the system consists of a recording device, an underwater sensor, and a cable to connect the two. In basic operation, the side scan sonar recorder charges capacitors in the towfish through the tow cable. On command from the recorder (set as a function of range) this stored power is dumped to the transducers, which emit the acoustic pulse that propagates out through the water. Then, over a very short period of time (from just a few milliseconds up to longer than one second), the returning echoes from the seafloor are received by the transducers, amplified on a time varied gain curve, and transmitted up the tow cable to the recorder. The recorder further processes these signals, digitizes them, calculates the proper position for them in the final record, pixel by pixel, and then prints these echoes on electro-sensitive or thermal paper one scan, or line, at a time.

The theory of operation of side scan is primarily concerned with the in-water end of the system, where much of the image forming process is controlled. Although the recorder is the command module for the system and contains the printer for the hard copy record, the transducers are the sound sources of any underwater acoustic system, and are key in the very beginning of image creation.

The active elements in side scan transducers are piezo-electric ceramic plates that expand or contract under the influence of an electric field. In order to construct a transducer array of the proper shape, numerous specially shaped ceramics or "crystals" are typically tied together electrically in a line. The array is then bonded to a reflective backing and encased in an acoustically matched compound. An electric field is applied by the triggering electronics of the sonar through thin layers of silver on each side of the ceramic plates. When voltage is applied to the plates they rapidly change size. It is this change in shape that provides the pressure wave to water in contact with the transducer face, thus starting the outgoing sonar pulse. After a number of cycles the ceramic array goes into a "resting" state. As the reflected echoes return from the seafloor in the form of low amplitude pressure waves, they strike the array and, in turn, cause a slight change in shape of the crystals. This change in shape is converted into an electric signal and transmitted up the tow cable.

Very careful dimensioning of the transducer array provides remarkably precise control of the shape of the acoustic beam (*beam forming*) which is transmitted from the sonar towfish. Side scan beams are very narrow in the horizontal plane. This allows the sonar to transmit across a very thin slice of ocean and seabed for each outgoing pulse. While it is narrow in the horizontal, this beam is also very wide in the vertical plane, allowing the sonar to transmit into the entire water column vertically. This fan shaped beam allows the system to record acoustic reflections from the surface of the water and anything in between the surface and the bottom.

After the acoustic energy reflected from bottom and waterborne discontinuities is received by the transducers and transmitted up the tow cable, it is printed on paper or displayed on a video monitor. What we see in the resulting sonar data is partly dependent on the beam form and its length (*pulse length*) as it propagates out from the towfish, as well as the form of the return echoes.

When a sound pulse is emitted from the transducers, it attenuates very rapidly. *Absorption* reduces the strength of the outgoing pulse and the returning echoes due to physical and chemical processes in the ocean. Absorption in the ocean is much more rapid than in fresh water. Absorption causes a linear reduction of echo amplitude with range when measured in deciBels (dB). This means that absorption, when coupled with other sound loss factors such as *beam spreading* and *scattering*, results in an exponential reduction of returned energy with range. Sound absorption (in dB/meter) increases roughly in proportion with frequency in the ranges commonly used by side scan sonar.

The sonar electronics must be able to display the returning echoes as though they are constant and even from the near ranges below the towfish out to the long ranges at the edge of the area scanned. This requires gain amplitudes to be increased rapidly after each outgoing pulse and immediately dropped back at the beginning of the next pulse. The first reflections from the sea floor near the towfish are at a very high level and the receiver gain in the towfish electronics is not required to be high. As the pulse travels out farther away from the towfish (usually in direct proportion to time), it becomes more and more attenuated, scattered and absorbed by the environment. Returning echoes from great distances are extremely low level and require very high amplification to normalize them. The means to provide this is typically based on time (given a reasonably constant speed of sound in water), and is known as *time varied gain* (TVG). Although the process is largely transparent to the sonar operator, time varied gain circuitry is crucial to quality data.

TVG in side scan receiving circuitry is the primary method used to maintain an even image across the sonar record, even though the returning signals have a far lower intensity at the outer limits of the range than in the areas close to the towfish.

TVG correctly assumes a reasonably constant speed of sound in water and an appropriate reduction in the returning echoes with range. The gain is increased along a complex, predetermined gain curve; although this is calculated by the sonar system, it is also partially controlled by the operator.

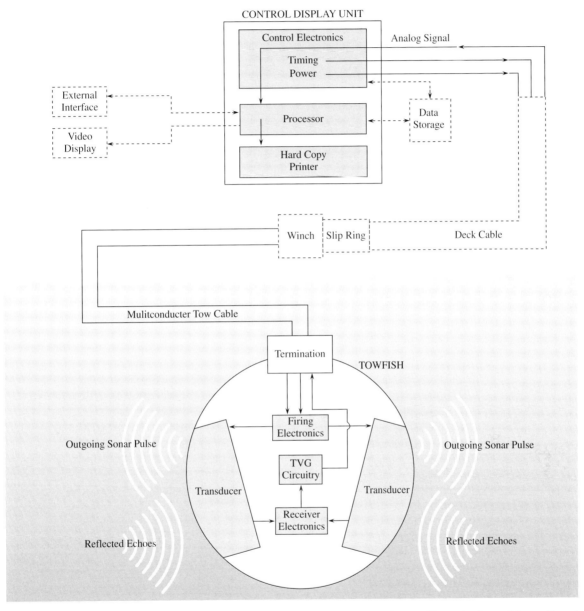

Illustration 12: In this block diagram of a side scan sonar system using multiconductor cable, optional components are outlined in dots. The control/display unit (recorder) contains the control electronics for timing the outgoing sonar pulse. The power and trigger pulse are transmitted down the tow cable. If a slip ring and/or a winch is used, an intermediate "deck cable" is used between the recorder and slip ring. In the towfish, the firing electronics cause the transducer to transmit the outgoing sonar pulse. The returning echoes from that pulse are received by the same transducer and sent to the receiving electronics. The signal is amplified and the time varied gain curve is applied. Some sonar systems apply TVG at the surface. The signal is then transmitted along dedicated conductors back to the recorder. Here the signal is digitized and processed for correction. It is then sent to the printer or video display. The data can be stored on magnetic tape or computer if desired. An optional interface to a navigation system will allow automatic recording of position and speed information.

Overall bottom reflectivity (known technically as *reverberation* under the influence of sonar") plays an important part in setting both gain and grazing angles in side scan sonars. Where a soft mud bottom reflects very little sound, a gravel or rock bottom is a good reflector and a mixture of these bottom types on the sonar record creates a challenge for the operator attempting to gain maximum information from the data. One of the methods that you may employ when encountering unusual bottom conditions is to change the grazing angle correction upon which the side scan system bases its time varied gain. This will often help to create more easily interpreted records and high quality data.

In the mid 1980's color video displays were developed for display of sonar data. These displays allow a greater dynamic range of detail than hard-copy paper. Further, since the image is digitized and the video display is controlled by software, video imagery can be enhanced in real-time, and the use of "windows" allows the operator to view several versions of the data simultaneously. When using a video display, the operator has the choice of running the hard-copy paper in the recorder, storing the data on tape, or both.

DISTORTION CORRECTED RECORDS

Side scan sonar data becomes distorted during generation. These distortions are caused by towfish instabilities (such as heave, pitch or yaw), speed variations in the survey vessel, and range data compression due to towfish altitude. Two of these distortions, those due to speed variations and range data compression, have been overcome in digital sonar equipment.

SPEED CORRECTION

Modern side scan sonar systems increase chart and printing speed to correlate directly with increased tow speeds. Known as *speed correction*, this means that the chart length over a given period of data collection will correlate directly with the ship's *over-the-ground* (OTG) speed. Early sonars simply increased the chart speed to compensate for higher tow speeds, but recent systems also write to the record at a higher rate, resulting in a far more readable record. They also have system interfaces which allow speed to be remotely input to the sonar system. In water where there is little current, or other water movement, an impeller connected to the sonar system is towed behind the boat. The impeller translates water movement into speed through the water to generate a speed figure. However, in most ocean environments, where tidal and longshore currents predominate, OTG speed should be remotely entered from a navigational instrument or calculated and manually entered into the sonar via the front panel controls.

Most navigational instruments, such as microwave, GPS or loran, will provide speed calculations to a sonar's navigational interface. However, some of these instruments, such as loran, are not fast enough to keep up with turns and other vessel course changes to provide the user with timely speed information. In these cases, it is often better for the user to enter the speed information manually. If this is done with reasonable accuracy, the sonar records will provide a good resemblance of the area scanned.

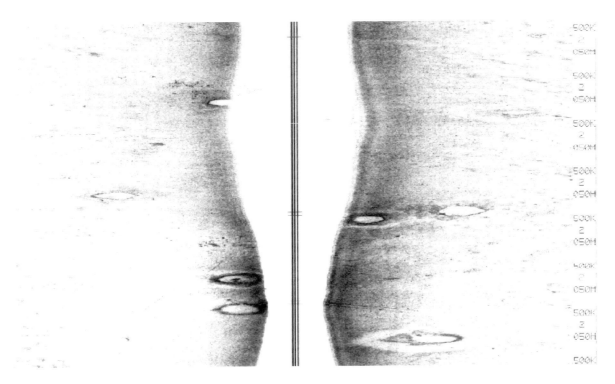

Record 13: This record of circular shapes on the seabed from detonated mines was generated using a much lower chart speed setting than the actual tow speed of the survey vessel. As a result, the image of the seafloor is severely compressed along the axis of the vessel towpath.

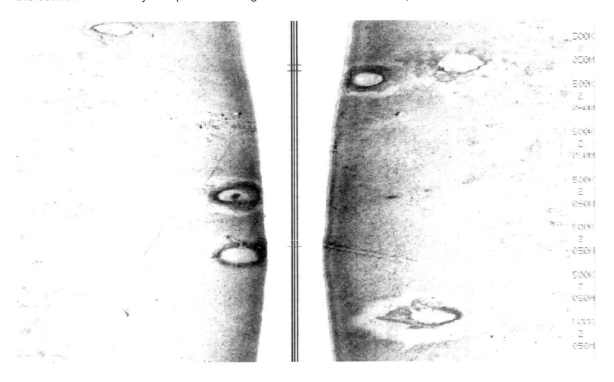

Record 14: A higher chart speed setting than that used in record 13 brings the image of the seabed closer to the actual bottom configuration, but some compression still occurs since the chart speed is not representative of the survey vessels speed.

Records 13, 14, 15, 16, 17 and 18 demonstrate the effects of speed distortion on sonar records. The first three of these records show an area of seabed upon which explosive mines were detonated. The resulting craters, which make solid sonar targets, are circular in shape. In Record 13 the speed data manually fed into the sonar was about 30 percent of the actual vessel OTG speed. The resulting sonar records show a severe compression of the circular targets in the direction of the towfish track. Record 14 was generated using a higher speed figure, but, as seen by the target compression, it is still too low. Record 15 was made with the proper speed figure entered into the sonar system. Although the shape of these mine craters is known, during most surveys, the operator would not have such benchmarks against which to gauge the accuracy of speed correction. Thus, reliable speed data must be entered during data generation.

Record 16 is a survey of a harbor area, again with a low speed value entered into the sonar system. Using speed figures matching the survey vessel's speed over the bottom, two sections are seen at their actual scale in the next two records. Record 17 is a footing at location "A" in the compressed record, and record 18 is a pier face at location "B" in the compressed record.

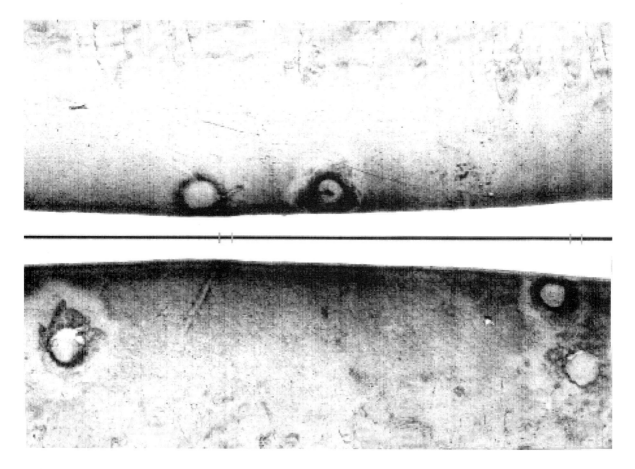

Record 15: With the actual speed of the survey vessel entered into the sonar system during record generation, along-track compression does not occur and the image is a fair representation of the overall seabed configuration in that dimension.

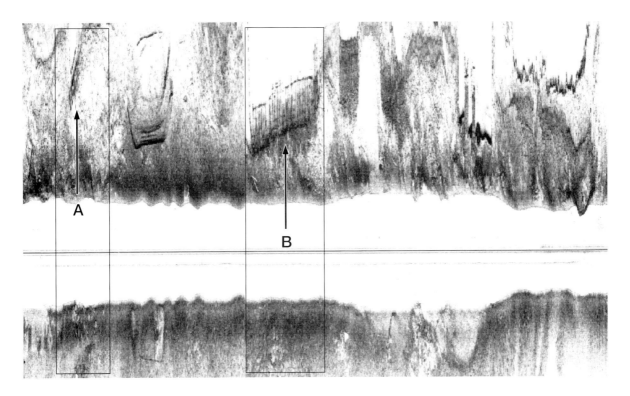

Record 16: In another comparison of speed distortion, low chart speed used results in severe image compression of man-made features in this record. (A) is a footing with detailed structure not visible here and (B) is a pier face at the base of which lie older discarded pilings.

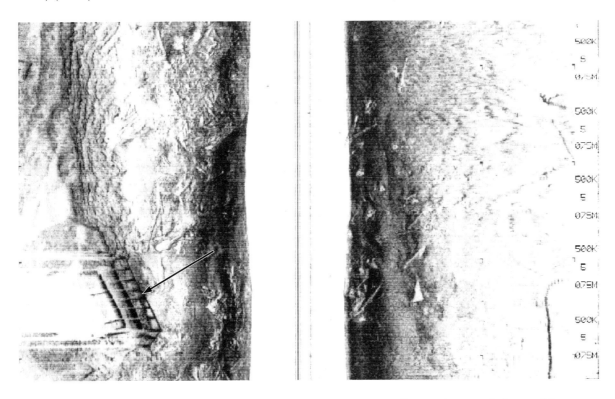

Record 17: Section A from Record 16 is shown generated with speed figures that match those of the survey vessel. The detail of the footing structure can be more clearly seen. (arrow)

Conversely, setting speeds that are too high will have the effect of elongating the record on the same axis that compression takes place in these examples. These records demonstrate that accurate record interpretation becomes significantly more difficult with non-speed-corrected records.

SLANT RANGE CORRECTION

Apart from the distortions induced into side scan sonar records by the speed of the survey vessel, there is another distortion caused by the actual physics of acoustic imaging called *range data compression*. Modern sonar systems also correct for this, but an understanding of how the data is distorted is required before the user can determine whether the data should be corrected for this type of distortion.

Although side scan sonar results appear to be an image taken from directly above the seabed (as in aerial photography), in reality, the initial return from the transducer is nearly vertical and the returns from long range are nearly horizontal. Each data point in between has some lateral, or range, distortion. Since one of the main functions of the sonar is to provide the operator with position information, range accuracy is important.

If the towfish were towed relatively close to the bottom of a flat seabed, the range information would be reasonably accurate since, for these sonar purposes, the speed of sound in water is reasonably constant. However, in practice, the fish is towed at some altitude above the bottom and this induces an error in true range on

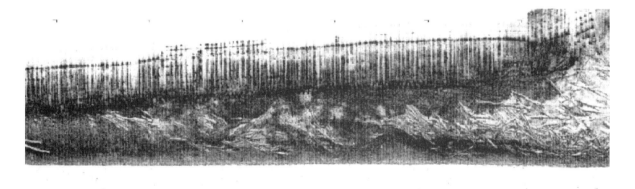

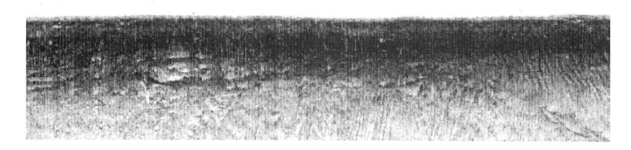

Record 18: Section B from Record 16 is shown with the proper speed settings. When speed settings are incorrect, only the along-track dimension is affected. Cross track (range) information on the records is based on the speed of sound rather than the speed of the vessel and is therefore unaffected by chart speed setting.

the horizontal plane. The actual range samples are along the hypotenuse of a triangle described by the towfish, its height taken above the seafloor at the horizontal range to the seabed point of interest. This *slant range* is the straight line distance from the transducer to the seabed point of interest. Slant range data, even with the chart speed corrected to match the survey vessel speed, does not give the operator a true 1:1 proportional image of the seabed. The greatest distortion occurs across track near the center, or near ranges, of the record.

Illustration 19 shows the reason why slant range data is distorted by towfish altitude and the time of returning signals. The distance (x) between targets A and B is the same as the distance between targets C and D. However, because the towfish is almost directly over the first two targets, the range to these targets is almost the same. As a consequence, the round trip travel for the sonar beam is almost the same. Since side scan sonar displays data based upon travel time, these two targets would be displayed closer together than they should be. Points at longer ranges from the towfish are displayed more accurately since the difference between the ranges to those targets is more representative of the distance between them.

Sonar systems designed and built in the 1960's and 1970's display the sonar data strictly as slant range from the transmission time up to the maximum range setting. This method puts the water column in the center of the record and provides the non-linear range increment discussed in the previous paragraph, across the display causing *slant range distortion*.

For many survey purposes, a true 1:1 linear representation is desired in the sonar data. With the advent of microprocessor based, digital sonar, systems are now

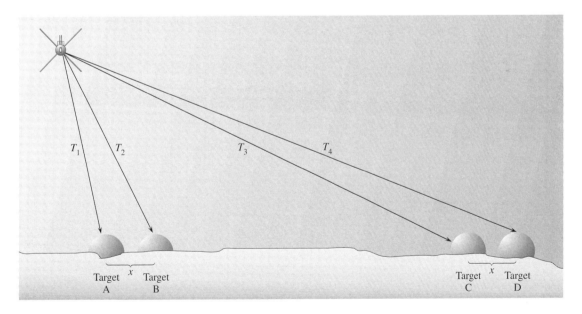

Illustration 19: Slant range distortion occurs because the imaging point (towfish) is positioned on a path above the seabed. Targets A and B are separated by the same distance as targets C and D. But because the range to A and B is almost the same (times T_1 and T_2 for the sonar beam), on a sonar record they will appear closer together than they are. Targets at longer ranges do not suffer as profoundly from this distortion as near range targets. The difference between sonar beam travel times T_3 and T_4 more closely represent the true ranges between these targets.

available that will remove the water column from the sonar record as well as provide slant range corrected data, pixel by pixel, across the entire displayed record. This correction is based on the altitude of the towfish and is done with high accuracy.

Since *slant range correction* is not required for all sonar operations, most systems now allow the user to choose either corrected or uncorrected operating modes. In an uncorrected sonar record, the outgoing sonar pulse propagates into the water and the first contact made with the seafloor is typically directly below the towfish as described in Chapter 4. The space printed on the record between the time of the outgoing pulse and the first bottom return is called the water column. The amount of space on the actual sonar record taken up by the water column is a direct function of the height of the towfish above the bottom. Disadvantages to printing the water column include using up record writing area and having to manually cut it out during the creation of mosaics. Advantages to printing the water column include getting data from objects that protrude off the seafloor. This is particularly valuable during fisheries applications, target search, and mapping and identification surveys. Whether or not the water column is removed or the sonar records are slant range corrected, depends upon the application of the system and the requirements of the user.

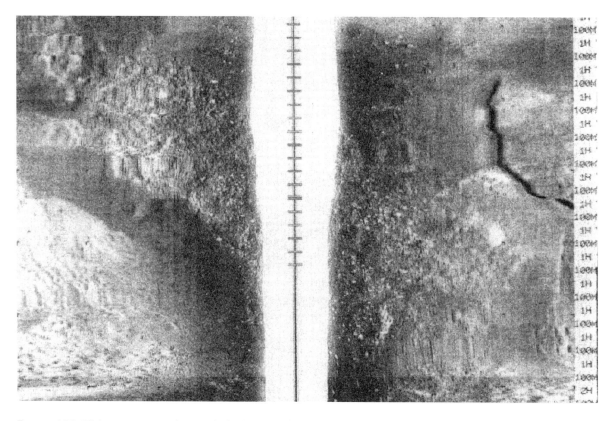

Record 20: This uncorrected record of a harbor floor shows a variety of sediment types. Small rock, ledge and boulders are shown crossing the record. A softer, fine sand shows up as a lighter area. The dark line on the starboard channel is an out-of-range return from a pier face 170 meters to starboard. In this record the water column is shown in the center and range compression is occurring in the areas close to the towfish. (compare with Record 21).

Record 20 is an image of a harbor seafloor with a mixture of sediments. This record is corrected for speed but uncorrected for slant range. The water column is printed and the sonar range is 100 meters. The tic marks in the center of the record indicate the towfish position at the time of navigational updates. The image in Record 21 is the same seabed but displayed in a slant range corrected mode and is a more accurate representation of the configuration of the area being surveyed. Note that the seabed nearer the towpath has been enlarged by the correction process as compared to the uncorrected record.

Record 22 shows an area containing both hard and soft sediments. The softer sediment is lighter in appearance in the record because of its lower reflectivity. In this record, the water column is printed and the fish is high enough to image the turbulence of the survey vessel's wake. Record 23 shows the same area with the data corrected for slant range. Here the water column (and associated turbulence) is removed, providing the user with a more accurate representation of the seafloor.

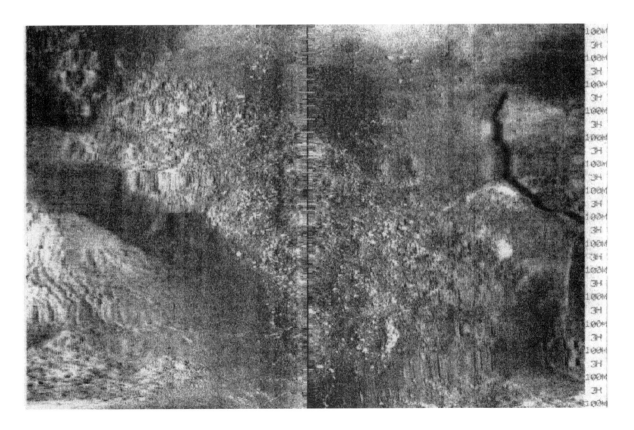

Record 21: The same area as shown in Record 20 using slant range correction feature of the sonar system. Note that the water column has been removed and the seafloor image in the near ranges is expanded. This represents the actual seafloor configuration more accurately than a non-corrected record. It also makes interpretation of topographic features easier. Corrected records are very important in creating seafloor mosaics.

Digitally corrected sonar data allows for the creation of mosaics of large areas of the seabed. When corrected, sonar records from a series of tracks are juxtaposed and matched, the resulting mosaic makes a comprehensive plan view of the seafloor. Such mosaics have contributed significantly to an understanding of the seafloor processes that affect coastal and deep ocean environments.

RESOLUTION

The resolving capability of any acoustic imaging system is an important feature for the surveyor. Realizing the capabilities and limitations of a system allows the user to put it to the best use. A number of factors must be considered when discussing the resolving power of the system. The ability to resolve multiple targets as distinct and separate on the seafloor, is a function of the sonar pulse width as well as beam spreading and the speed of the towfish through the water.

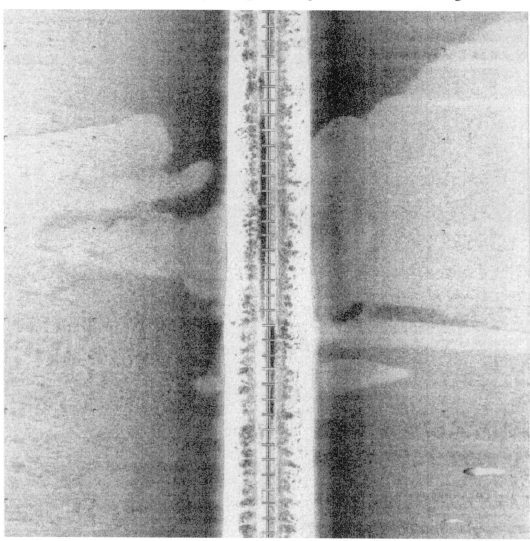

Record 22: This uncorrected record shows areas of hard sediment mixed with patches of softer seafloor material. Again, the uncorrected record contains the water column and range compression occurs. Note the reflections of the survey vessel wake in the water column section of the record (compare with Record 23).

The two types of resolution about which we need to be concerned are *transverse resolution* (resolving similar objects that lie in a line parallel to the towpath) and *range resolution* (resolving small separate objects that lie in a line 90° to the towpath). These are formally defined by Flemming et al. (1976) as follows:

Transverse resolution is the minimum distance between two objects parallel to the line of travel that will be displayed on the sonar as separate objects. This minimum distance is equivalent to the beamwidth (which widens with distance from the towfish) at any particular point. Range resolution is the minimum distance between two objects perpendicular to the line of travel that will be displayed as separate objects. This is a function of the display system as well as the topography but more exactly, the pulse width of the sonar will determine the lower limit of detecting the objects as distinctly separate.

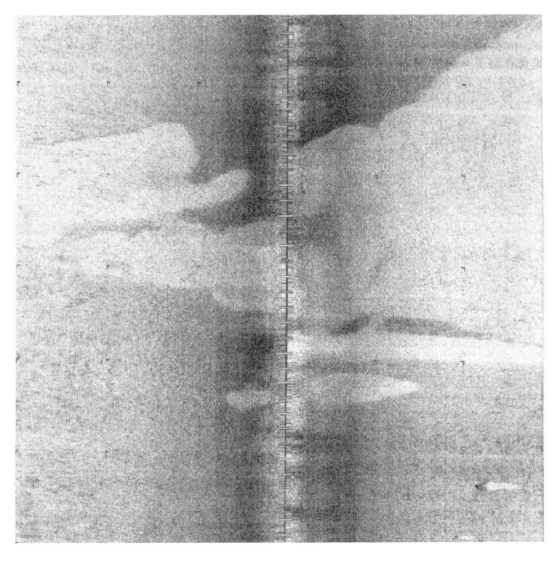

Record 23: The same area of seabed depicted in Record 22 is shown in the slant range corrected mode. This removes the water column along with the image of the survey vessel's wake and brings the two first bottom returns together while removing the range data compression from the record.

Transverse resolution, or that ability to discern two separate objects that lay near one another in a line parallel to the tow path, is a function of vessel speed, ping rate (range setting) and, most importantly, the horizontal beam spread of the sonar. Whereas tow speed and ping rates are controlled by the operator and *beam spreading* is not, it is important to understand the effects of beam spreading. Near the towfish, where beam spreading is not significant, two objects are clearly delineated as separate and distinct targets. However, as the beam spreads out into the far region, the areas of insonification widen for each outgoing pulse. In these cases two separate targets on the seabed may be imaged by the same sonar beam, thus appearing as one single target in the resulting sonar data. Bringing these targets into the near ranges on successive passes will resolve them as distinct targets. Illustration 24 shows this effect by using an exaggerated beam angle for clarity. Beam spreading also has the effect of degrading the overall resolution of single complex targets (see Records 72 and 73).

Range resolution is determined by knowing the pulse length of the sonar and understanding what happens to the sonic footprint (the seabed area insonified at any single point in time) as it propagates out across the seafloor. If the pulse length, translated into a section of water, is .15 meters in thickness (pulse length multiplied by the speed of sound in water), it could not resolve the difference between two objects that lay .05 meters apart. In this case the pulse would be

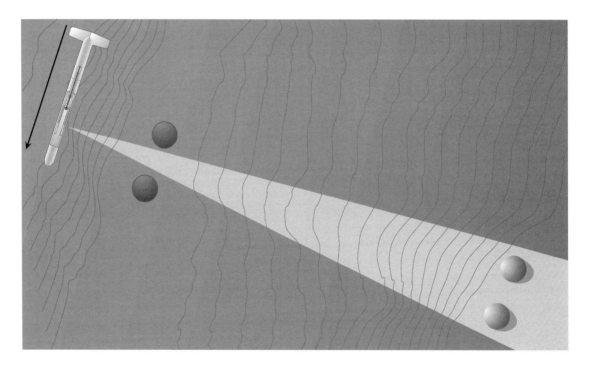

Illustration 24: Beam spreading effects the transverse resolving power of side scan sonar. In the near regions, where the beam is narrow, two closely spaced targets will be insonified and imaged as separate and distinct objects. As the beam widens, propagating away from the towfish, the same beam will insonify two closely spaced targets and they will appear as one single target on the sonar record.

enveloping both targets at the same time, therefore the returning echo would resemble a single target. Thus, a smaller pulse length (often used in high frequency side scan) will help generate a record with higher range resolution (see Illustration 25).

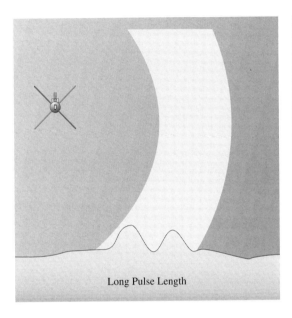 

Illustration 25: The pulse width of the acoustic beam will effect the range resolution of side scan sonar. A shorter pulse width will first insonify one target, travel beyond it, then insonify another close target. The record will show two distinct targets. A longer pulse width will encompass both targets at a given point in time and they will appear as one target in the record. Since the pulse length can be translated into distance in water, the operator can determine whether two targets with a known separation can be resolved.

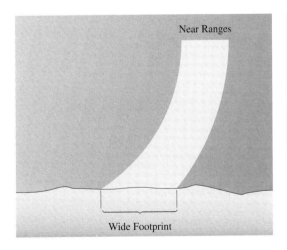 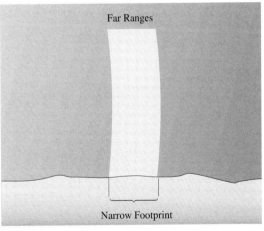

Illustration 26: Although beam spreading reduces resolution as sound propagates further from the towfish, range resolution increases with propagation. The outgoing sonar pulse forms an arc in the water. Where the beam strikes the seafloor in the near ranges, it's footprint is larger than when it is further away from the towfish. The smaller footprint will resolve two objects that are closer together.

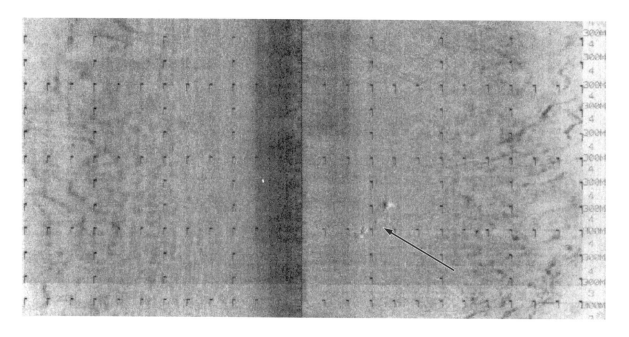

Record 27: This record, made using a range of 300 meters per side ,shows a badly deteriorated shipwreck. The wreck is 50 meters in length but the long range setting of the sonar results in a reduced image size.

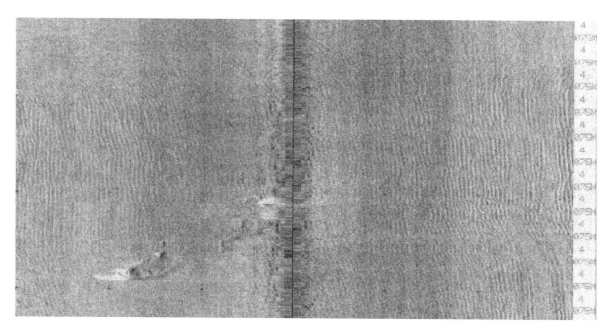

Record 28: The same shipwreck shown in Record 27 is scanned here using a range setting of 75 meters per side. The image of the ship's remains is much larger and is the dominant feature of the record. Sonar ranges should be chosen carefully when performing either search or survey tasks to provide target image sizes that are easily recognizable.

The range resolving power of any frequency of side scan sonar pulse also changes during each sweep. This occurs because as the beam propagates away from the transducer, the footprint of the insonified region of the seafloor changes. At the closer ranges, and particularly under the towfish, the beam's arc creates a larger footprint than at the longer ranges. Away from the towfish, the beam forms a greater arc and therefore a smaller footprint (see Illustration 26).

RANGE SETTINGS

Side scan sonar ranges refer to the *range* of the display on each side. For example, a range setting of 150 meters provides a total swath width of 300 meters. Resolution in any sonar system is not only a function of the sonar beam, but it is also a function of the display mechanism. When using a video display or a paper recorder, it is important to keep in mind that a constant sized display has a limit to pixel size and that shorter ranges provide higher displayed resolution. The resolution of each pixel is a function of the range divided by the number of pixels in the display that are spread across that range. During general area surveys, longer ranges give greater coverage, thus reducing ship time. On the other hand, shorter ranges provide higher levels of small object recognizability. Record 27 shows a large but heavily deteriorated shipwreck detected at the 300 meters per side range setting. It is recognizable as a discrete target, however, Record 28 of the same shipwreck scanned on a range of 75 meters per side provides a far greater recognizability. In determining the range for any particular operation the surveyor should consider the total area to be covered, and the minimum size of features to be resolved and recognized.

Due to signal losses from absorption, scattering and beam spreading, side scan sonar has a limited practical range. The exact distance of these range limits depends on the application. For instance, large reflectors can be detected at longer

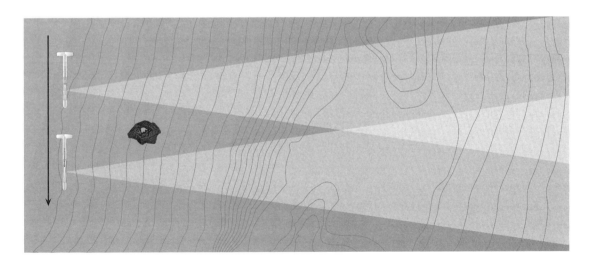

Illustration 29: The sonar emits carefully timed pulses as it is towed forward over the seabed. The time between two successive pulses is determined by the range setting of the recorder. On long range settings this can be as long as four to eight tenths of a second, and at high tow speeds, the fish can travel considerable distances between outgoing sonar pulses. Although beam spreading increases seabed coverage at long ranges, high tow speeds, combined with long range settings, can prevent the insonification of small targets.

ranges, but subtle differences in bottom topography may only be detected at shorter ranges. Thus, for large object search operations, the usable ranges might be twice that used for topographical surveys.

In fresh water the usable ranges for all sonar applications are greater because of the makeup of the medium. Fresh water contains much less of the chemical magnesium sulfate and other salts that are largely responsible for sound absorption in the lower side scan sonar frequency ranges. The elastic properties of these chemicals under the influence of sound energy, results in sonar energy losses. Where a high frequency (500 kHz) side scan system provides usable data to 130 meters in seawater, the same system could provide similar records to 165 meters in fresh water. Similarly, the medium frequency (100 kHz) is usable for general surveys to over 600 meters in sea water, but it might provide a usable range of over 1000 meters in fresh water. However, these are upper limits for conventional sonar systems, and for most practical purposes, shorter ranges are used. The optimal range for any application must be determined by the user to include a recognizable display of any needed data, while effectively utilizing ship and personnel time.

Also important for both search and survey operations are records of a quality that allow the operator to recognize targets.

SEABED ANOMALY ACQUISITION

DETECTION

The first step toward operator recognition of a target on the seafloor is detection, i.e. the sonar system must detect the seabed feature or anomaly. There are targets too small to be detected by conventional sonar, and there are operational scenarios where the system does not detect the target due to improper application of the equipment or because of noise in the environment. By examining the acoustic and mechanical design of the sonar system, the operator can determine if the feature can be detected. If the target is large enough to be detected, the user must be sure that the system is applied in such a manner to assure that it will insonify the target with the sonar beam. Along-track insonification, or the ability to actually insonify and record a target as data, is a function of both beam spreading and tow vessel speed over the bottom.

INSONIFICATION

A range setting of 750 meters per side will cause the sonar system to ping at a rate of 1 ping per second (assuming a sound speed of 1500 meters per second in sea water, the sonar will require 1 full second to transmit and receive the two way travel of the acoustic pulse at the limit of the range). If the survey vessel is traveling at a rate of 2 meters per second, in the near ranges where beam spreading is still narrow, the sonar beam may not insonify an object that is 1.5 meters in size. In the far ranges, this would not be a problem due to the overlap of consecutive beams, but for near-region targets, especially when the towfish is flown low, the operator should be aware of the speed versus range relationship (see Illustration 29). This is particularly true in target search operations because the range of target contact from the towpath is almost never known before the target is located. To carry this to extremes, if a search operation uses a vessel speed of 5 knots (approximately

2.5 meters per second) and a range of 600 meters per side, the survey may completely miss a target with a least dimension of 2 meters (6.5 ft). In this case, if the target were close to the towfish, where the sonar beam is at its narrowest, the target may not be detected even by one sonar ping.

In Records 30, 31 and 32, a seabed target is scanned at different tow speeds and sonar range settings. The target is a shipwreck. The survey track lines are at 90° to the target's long axis; this is a worst-case scenario since the target presents the smallest area for insonification. In Record 30, the target is insonified by only 2 or 3 sonar pulses. The range setting is 200 meters per side and the tow speeds are in excess of 6 knots. The target is detected, but it is barely recognizable in the sonar record. Given the potential for other variations in the topography and possible water borne discontinuities, a sonar operator may miss this as a viable target. Certainly, if the sonar ranges were set longer (lower ping rates) this target may not have been insonified by even one sonar pulse and therefore not detected, displayed, or recognized at all.

RECOGNITION

Recognition is the last step in the process of acquisition of a seabed feature or target and it is most dependent upon the eyes and brain of the operator. It is at this stage that an experienced operator may recognize a target missed by another operator. For instance, a man-made debris field mixed into a rocky bottom will certainly be detected and highly insonified, but the operator's eye might not distinguish between the natural and man made image components (see Records 64 and 65).

Record 31 shows the same target in Record 30 but scanned at a speed of about three knots. This slower speed allows more insonifications of the target and a more recognizable display. The target is also at a lesser angle to the towpath and this provides a more favorable profile for insonification, further enhancing the image. Record 32 shows the same target scanned at a lower range setting (higher ping rate) providing an even more recognizable image to the operator.

Even once detected by the sonar system and displayed on the resulting record, the target may not be recognized unless enough insonifications and an appropriate chart speed, allow it to be displayed in enough consecutive scans for the eye and brain of the operator to recognize it as a target. Although recent developments in side scan sonar have provided for repeated data displays in between outgoing pulses, a conservative plan includes one which chooses tow speeds and ranges that will allow for at least 12 insonifications (pings) in a forward travel distance equal to the targets least dimension (see Records 156 and 157). It is important to calculate maximum tow speeds and maximum ranges for target search in programs involving large search areas. The speed table in Appendix A can be useful in these calculations.

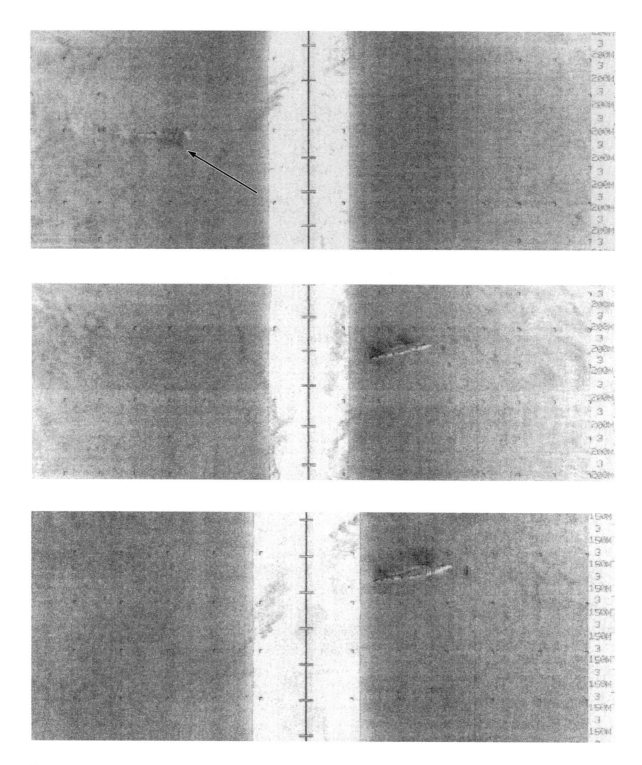

Records 30, 31, 32 : In Record 30 a medium sized target was scanned using a range setting of 200 meters per side and a high tow speed (>6 knots). The target, a sunken submarine (arrow), is insonified with only one or two sonar pulses because of the low ping rate and high tow speed. The target's long axis lays 90° to the survey vessel track making detection even less likely. If the target were further away and within range it might have been better detected because of beam spreading. For Record 31, the tow speed was reduced and the target was scanned at a slightly different angle, presenting a better profile for imaging. In Record 32 the range was reduced, thus increasing the sonar ping rate, and insonification.

Chapter 6:
System Configurations

Side scan sonar systems come in a variety of types, depth ratings, and frequencies. There are systems designed for use in ocean depths of 35,000 feet and there are systems designed for use only in the relatively benign environment of lakes and rivers. However, all systems require care in application to efficiently gain the highest resolution, and accordingly, the greatest information possible in the data gathered.

In the past two decades, underwater imaging has become highly refined by the use of the digital, microprocessor based system like those described in the last chapter. The standard system in use today is a portable, high-resolution system that relies on one of two or three frequencies for imaging. As seen in the previous pages, these frequencies are used to transmit a carefully formed beam and receive the returning echoes. Future side scan systems will include phased arrays, interferometric sonars, and chirp or sweep sonars that are effective in specialized applications. However, for the majority of underwater imaging operations, the multi-frequency, medium range side scan sonar of 50 to 500 kHz will provide the required data, easily and cost effectively.

Today's sonars are available in different frequencies, each for use in different applications. Lower frequencies such as 25, 27 or 50 kHz will propagate farther underwater, covering a larger area, but due to the longer wavelengths, they generate a lower resolution image. As a result, although they provide a swath width of 5 kilometers or more, these lower frequencies are only used to image large targets or cover wide areas. In contrast, a higher frequency, 300-500 kHz sonar, will resolve objects just a few inches in size. These high resolution systems only cover an area of a few hundred meters in width and so are usually used to survey fine details in small areas, as in minehunting operations, or search for heavily deteriorated target sites like very old shipwrecks. As a compromise between optimum range and high resolution, the medium frequency side scan sonar of 90-125 kHz is useful for a majority of imaging applications.

Most of the modern day sonar systems are designed to be portable and quickly mobilized to vessels of almost any size. Three components make up the basic system commonly used for underwater search or survey: a control/display unit (recorder/processor), a towed transducer assembly (towfish), and an electrome-chanical cable which connects the two.

RECORDER

The control/display unit, which produces the graphic records, contains tuning controls for the system. Historically, most conventional sonars also included the hard copy printing mechanism in this component of the system. For this reason, the control/display unit is often referred to as the "recorder". The recorder provides the operator with controls required for operating the entire system. Along with basic controls, it allows the user to choose gain levels, frequency of the sonar, chart speed, and slant range correction for survey operations.

Recorders are equipped with either electro-sensitive printing heads or thermal printers. On well designed systems the print head is rugged, provides a long print life, and is protected from the environment found at sea. The recorder also has a variety of interfaces for keyboard and navigational inputs, video displays, data storage and computer processing. As a basic adjunct to the recorder, color video displays with target analysis modules are very helpful in some applications, and using a video system during search often eliminates the need to run a hard-copy paper recorder until a target is located. Tape or disk recording of data is helpful if further processing of the data is desired. However, recorded data is crucial for the formation of mosaics (see Chapter 9).

As important a system component as the recorder is, it must receive the properly processed signals in order to produce a clear, well balanced sonar record. It is this data that is delivered via the tow cable from the towfish. The recorder not only provides the "number crunching" power required for microprocessor control of slant range correction, but also further processes input data from the towfish in terms of matching the signal's dynamic range to that of the display. The data from the towfish varies in amplitude due to four basic factors: range, transducer beam pattern, grazing angle of sonar with seafloor, and the seafloor and target *reverberation strength* (roughness). The signal input to the towfish covers a very wide dynamic range. One of the major advances in sonar technology has been the development of systems that process and accurately display this wide range of signals.

Overall, the side scan recording unit is the command component of the sonar system. It allows the operator to easily make either corrected or uncorrected images. At the same time, if the system is used in unique applications where the image requires modification, the recorder accepts fine tuning adjustments as desired by the operator.

TOWFISH

The component containing the transducers and signal conditioning circuitry is the submerged towfish. Towing the transducer array instead of hull-mounting it keeps the system portable, and most importantly, decouples the towfish from the ship's motion. This provides a stable imaging point, which is crucial to high quality data. Hull mounted imaging sonars usually require very sophisticated roll, pitch and heave sensors interfaced with a computer system to process the data and remove ship-motion-induced data errors. Towing the transducer array also places it nearer the bottom which is desirable for quality sonar data.

The towfish used for continental depth sonar surveys is a compact metal cylinder which can be handled by one person. Deeper versions are commonly heavier and require more significant deck handling equipment to deploy. This light weight towfish used in most systems operates on one of two frequencies. A high frequency for detail imaging and a lower frequency for long range searching.

These frequencies are switch selectable from the surface unit. This eliminates the need for two towfish, allowing for the identification of targets at long range with the lower operating frequency and then higher resolution classification of the same targets with the higher operating frequency. The towfish is equipped with

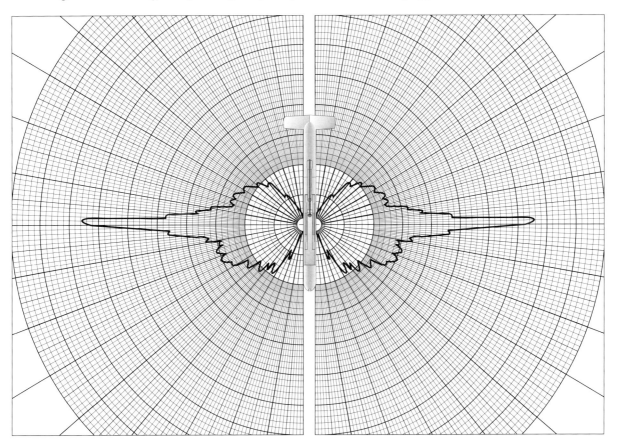

Illustration 34: The side scan sonar beam is carefully formed to provide a high resolution image of the seafloor and small targets. As viewed from above, the sonar beam has the highest energy level in a very narrow section. This is the main lobe of the sonar pulse that propagates out across the seabed. In a well designed transducer for side scan, side lobe energy is minimized.

stabilizing fins and ballast for hydrodynamic stability. Identical but separate port and starboard transducers are located on each side of the towfish. The electronics pressure housing section contains two printed circuit boards, one each for the port and starboard channels. Each board contains transducer driver and amplifier circuitry.

Some towfish incorporate a breakaway shear pin for use in the event of a collision with an object on the seabed. With this shear pin, if cable tension on the towing bridle exceeds a specific value due to snagging on the seafloor or on an obstruction, the pin shears, and the electrical connection between the cable and towfish separate. The towfish then turns 360 degrees, transferring the tow point first aft and then to the nose of the fish allowing it to be pulled clear of the obstruction. The fish can then be recovered, rearmed and redeployed. If the tail fins snag, they are designed to fall away, but remain attached to the towfish by a cord. If the snag persists, the cord will break to avoid loss of the towfish.

The maximum tow speed for generating records is dependent upon the application, but it is generally limited by the hydrodynamic drag forces on the towfish and cable. At the upper limit, these forces will bring the towing assembly to an unstable position at the surface. Depending upon the towing configuration this will occur at speeds above 12 knots.

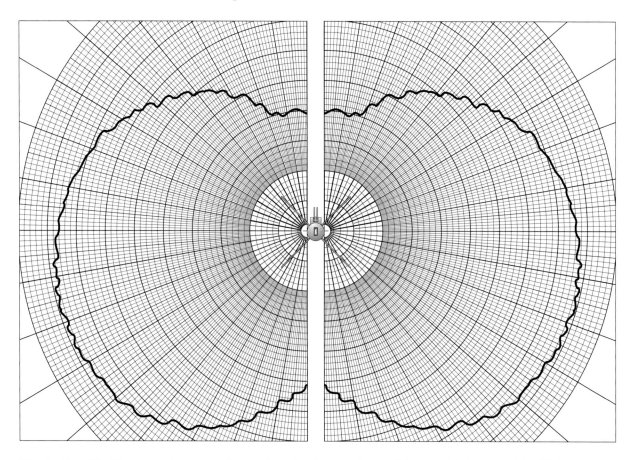

Illustration 35: The sonar beam as viewed from the front or back of the towfish is very wide. This provides near and far seafloor coverage. In the sonar towfish this beam is tilted downwards between 10 and 20 degrees.

The outgoing pulse length, or amount of time the transducer transmits sound, is typically 0.1 millisecond for the medium frequency pulse and .01 millisecond for the high frequency pulse. Both the shorter wavelength and the shorter pulse length account for the better resolution of the high frequency sonar.

These carefully formed sonar beams are dimensioned to generate sound waves shaped specifically for generating the highest resolution possible with the given frequencies. With the acoustic pulse transmitted and received from the same set of transducers, the *horizontal beam angle* is approximately 0.6 degrees for the medium frequency beam and approximately 0.25 degrees for the high frequency. The overall *vertical beam angle* is 50 degrees. The transducers are usually tilted down from the vertical and the vertical beamwidth is independent of the downward tilt whether it is 10, 15, or 20 degrees. A representation of typical side scan sonar beamwidths in both the horizontal and vertical are shown in Illustrations 34 and 35.

FREQUENCY DIFFERENCES

The two common side scan sonar frequencies used for continental shelf depth applications are referred to as 100 kHz and 500 kHz; however, these frequency labels are generic. In actuality, the frequencies used are typically within 20% of these figures depending upon the system's manufacturer. For the sake of

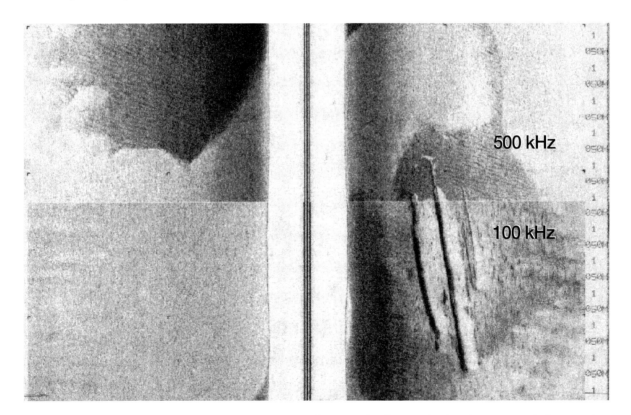

Record 36: The effect of the two primary side scan frequencies is seen in the above record of a wooden shipwreck. The wreck is lying on a seabed of silt (lighter sediment) and sand. The operator switched from 100 kHz to 500 kHz after most of the target was scanned at the lower frequency. The better definition the high frequency furnishes is due to the shorter wavelength and the shorter pulse length.

simplicity in this text, the medium and high frequencies will be referred to as 100 and 500 kHz, respectively. Most well designed sonar systems incorporate both frequencies. The 100 kHz is used out to 600 meters for a large target, or general area search. The 500 kHz, on the other hand, provides a higher resolution and usable ranges extend out to about 100 meters although target detectability occurs as far as 150 meters under ideal environmental conditions and proper sonar geometry. The differences between the two frequencies gives the user the advantage of a long range sonar and a high resolution sonar in one instrument. In Record 36 the operator switched from 100 kHz to 500 kHz while scanning the remains of an old shipwreck. The near range clarity of the image generated using the higher frequency is seen in the resulting data.

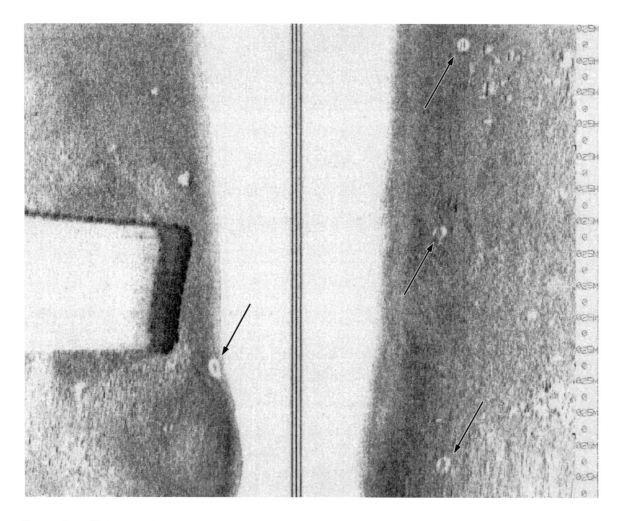

Record 37: This record was generated using 500 kHz at a marine construction site. There are several rubber truck tires lying on the bottom. Rubber, along with some woods and plastics absorb the higher frequencies and return very little energy to the transducer. Anechoic tiles placed on a subsea target will have the same effect with a variety of sonar frequencies. Note that the targets, as well as their associated shadows are clear on the record. The elongation of the tire on the port channel in front of the cassion is due to range data compression and it has little shadow because of the sonar geometry.

The records provided by the 500 and 100 kHz differ in other ways as well. Substances found in the underwater environment reverberate differently with the influence of sound. In addition to the increased resolution of the 500 kHz record, the operator may find that some objects seem to absorb 500 kHz sound waves with little reverberation; this causes the objects to disappear in the sonar record background leaving only a light trace of their existence along with their normal shadow. Natural objects in this category include the lighter woods such as soft pine and balsa and some rubber compounds. Man-made objects in this category include anechoic targets coated with acoustic tiles. These anechoic targets are specifically designed to be invisible to sonar. In Record 37, made near an underwater construction site, four large rubber truck tires lie on the bottom. The beam of the 500 kHz sonar was not reflected off the tires and did not cause an appreciable amount of reverberation. As a result the operator infers their existence by the lack of returning signal.

CABLES AND CONNECTORS

The towfish is connected to the tow cable by an underwater (wet) connector for electrical connections, and to the towing bridal for the mechanical connection to the survey vessel and the recorder. A reinforced lightweight multiconductor cable for shallow operations, or an armored one for deeper towing are the most common cable configurations. For very deep operations, a single conductor, armored cable is used. The internal makeup of these cables depends upon the depth and application of the system.

In near-shore waters ranging from just a few meters to 75 meters in depth, the light weight cable is used. These cables vary in length from "test cables" of 10 meters to cables 150 meters in length. The light weight cable is favored because it is easy to handle on deck and can be transported by one person. It has either an Aramid or metal strength member built into it. The working load of 1.3 centimeter light weight cables is between 400 and 1000 kilograms, depending upon the size of the strength member inside. The in-water weight of these cables is also dependent upon the internal makeup of the cable, but typically these cables are very light in water. Unfortunately, because of this, these flexible and easily handled cables do not contribute to the effort to "depress" the towfish closer to the bottom.

Light weight cables are manufactured with a waterproof jacket. In reality, these jackets are violated fairly easily in use. The primary protection for the conductors against water contact is the conductor insulation itself. Most cable manufacturers agree that this is well within the design requirements of the cable.

Armored cables are used for deeper (>75 meter water depth) towing. They are most easily handled on a winch system and hold up well under the stresses of large depressors. Most common armored side scan cables are "torque balanced" and "double-armored," meaning that they have two layers of contrahelically wound steel jackets. Many are self flooding and rely on the conductor insulation to protect the electrical conductors against water contact. The standard armored multiconductor side scan cable is less than 1.3 cm in diameter and is several times heavier in water than its light weight counterpart. It has a working load of between 4000 and 7000 kilograms. For longer lengths of armored cable an electric or hydraulic winch system is highly recommended. The accessories to the winch include a meter wheel to measure the amount of cable out, a tension meter to measure the force on the winch and cable, slip ring (see below), and level wind to maintain layered cable on the winch drum. Multiconductor side scan cables used in most continental shelf depth operations have between 7 and 9 conductors, over which power, trigger, ground, and returning sonar data are all transmitted between the towfish and the recorder.

The dry connector on the recorder end of the cable and the wet connector on the towfish end of the cable are delicate and should always be handled with care. The connectors are the most frequent area of tow cable failure and should be the first components checked on a cable suspected of being faulty.

SLIP RINGS

Slip rings are useful if a cable handling winch is employed. Slip rings are an electromechanical component that allow, through the use of motor-type brushes, full time electrical continuity between one section of cable under rotary motion (such as wire on a winch drum) and a stationary section of cable (such as that connected to the recorder). A slip ring permits the continuous operation of an electromechanical cable during winch drum operation. The slip ring is mounted on the winch and is connected on one side to the towing cable. The other side of the slip ring is connected to a simple deck cable and the recorder. Between these two connections, movable brush-like contacts provide full time electrical continuity from the towfish to the recorder.

Some slip rings contribute considerable noise to sonar systems, particularly during ring rotation. Slip rings may require cleaning, brush replacement or other maintenance periodically. Since they become part of the conducting components of the towcable, they are often suspect in a faulty cable assembly mounted on a winch.

DEPRESSORS

When towing side scan sonar with short lengths of in-water cable (<100 meters), the weight of a standard towfish is usually sufficient to maintain a fish depth of about 30-40 meters, especially if tow speeds are 3 knots or less. When tow speeds are higher or when in-water cable lengths become greater, the drag of the cable in the water column overcomes gravitational force on the tow assembly. This has the net result of raising the towfish away from the bottom.

In shallow water operations using a cable and towfish alone, tow speeds affect fish height by changing the drag force on the cable. Because of this increased drag, increasing tow speeds can raise the towfish thus help to avoid collisions with the bottom. At high tow speeds these drag forces will bring the towfish to within a few meters of the water surface even with hundreds of meters of cable in the water.

One may think that simply deploying more cable will lower the fish in deep water. However, with sufficient cable payed out at a given tow speed, more cable may not increase the depth of the fish because the added cable only increases drag, keeping the towfish high in the water column. In this case, the cable fed out usually has very little angle into the water and tows straight out along the towpath. The use of the heavier armored cable will add weight to the in water components and help bring the towfish closer to the bottom. When working with higher tow speeds and/or deeper water, a *depressor* is required.

A depressor is a mechanical component added to the towing assembly near the towfish that aids in overcoming the effects of cable drag by bringing the towfish deeper. Most depressors are added to the cable ahead of the towfish by 10 or 20 meters. This will minimize any negative effect that may be caused by any slight instability (poor towing behavior) in the depressor.

One way to depress the towfish is to add weight to the cable in the region of the towfish. Known as *deadweight depressors*, these weights are inexpensive, can be constructed from a variety of materials, and are easily formed into hydrodynamic shapes so that they tow well. Disadvantages to the deadweight depressor are that it requires heavier deck handling equipment for deployment and recovery, and its effectiveness is not enhanced at increased speeds.

Hydrodynamic depressors are also used to add depressive forces to the towcable. Hydrodynamic depressors utilize the towing action of the survey vessel to increase the downward pull on the cable. These accessories have one or more vanes or louvers which, when tilted down at a fixed angle, force the depressor down when it is pulled through the water. The hydrodynamic depressor is analogous to an inverted airplane wing. One advantage to the hydrodynamic depressor is that it is very light and does not require heavy deck handling equipment for deployment and recovery. A disadvantage to this type of depressor is that when the surveyor needs to rapidly decrease the depth of the fish to avoid collision with the bottom, increasing vessel speed may have the opposite effect by pulling the fish deeper rather than shallower. Another disadvantage to the hydrodynamic depressor is that cable tension will increase proportionately with increases in tow speed.

Light weight cables are not typically used with depressors because these cables are for shallow water use. Light weight cables also have a far lower breaking strength than armored cable and cannot accept the considerable tension added by a depressor. In most applications, the first step in depressing a towfish is to use an armored cable. Depressors can be mounted onto the towfish itself, although for medium and deep operations the depressive forces desired are usually only gained from mechanical attachments to the cable using line grips. Line grips are often helically wound accessories that attach to the cable. Some grips must be attached before the cable is terminated while others may be attached at any time without disturbing the termination. They are manufactured in a variety of materials including steel and Aramid fiber. American manufacturers of cable grips include Kellums, Yale, and Preformed Marine.

Any deployment of a towing assembly using a depressor should be done with some amount of way on the survey vessel, otherwise, with the towing assembly hanging vertically in the water column, the bridle or other attachments of the depressor could twist around the tow cable. This action would add undesired torque to the lower portion of the cable thereby affecting the stability of the towfish, during towing.

NAVIGATION

One of the most important pieces of equipment for either general area survey or specific target imaging is a navigation system. A navigation system is defined as any method or equipment which allows the surveyor to: **1.** be able to travel between any two points on the surface of the water along a predetermined track line, **2.** know precisely where he is at all times, and **3.** be able to return to any point at a later time. These systems and methods range from using simple shore ranges to loran or sophisticated satellite based positioning systems. Unlike magnetometers and bathymetric or subbottom profilers, which are limited to narrow swaths, the coverage of side scan sonar allows the use of navigation systems with tolerances that approach 100 feet or more depending on the necessary sonar search range used.

Although this text is not written with the purpose of outlining the possible configurations of the many different navigation systems available, some basic information is helpful in planning sonar surveys.

In near shore areas, and during a survey where track lines must be followed very closely, a simple but manpower intensive system has been used very successfully by the authors. This procedure, used often by the late Dr. Harold Edgerton for shallow water work, consists of using two or more range stakes set on shore at a specific distance apart (in line with the vessel track). If the separation of the range stakes is sufficient, the pilot can run a track line very accurately while keeping the stakes in line. For subsequent track lines, the shore team is required to measure off the distance to each successive stake location and to position it so that each track is equidistant from and parallel to its predecessor. This is best done before the survey starts in order not to delay the offshore crew needlessly. Although this system of track line navigation may seem crude, for near shore survey work it is very accurate and is seldom surpassed by even very sophisticated electronic equipment. Unfortunately, most surveys are not performed within areas that are close enough to shore to use the range stake method. When performing these near shore surveys, it is important not to use single point, or any kind of radial search pattern since they represent little more than random search methods and have a great potential for "holes" or gaps in the search area. However, the range stake method has proved to be one of the few accurate methods of manual navigation.

For operations that occur farther offshore, microwave, high frequency, or laser based systems will provide meter-accuracy for the surveyor; although in many cases, particularly for target search operations operating at long ranges, loran is sufficient to effectively cover large areas. More and more operations deploying side scan sonar are relying on *GPS* (Global Positioning System), a Doppler satellite system, for accurate navigation. Since both the loran and GPS systems are somewhat dependent upon atmospheric conditions for accuracy, they are not as accurate as the shorter range microwave or laser systems. However, by setting up a shore station at a surveyed, "known" position and transmitting this data to the survey vessel, the survey accuracy based on loran and GPS is increased significantly. This process, known as *differentiation*, is commonly used to gain accuracies of a few meters with these navigation systems. As efficient as these differentiated systems are for most coastal side scan applications, standard loran or GPS will provide the navigation data required.

As discussed, navigation is one of the most important components of any search or survey. Accurate navigation not only allows the surveyor to cover a predetermined area without excessive repetition or gaps in coverage, but also allows him to return to any areas of interest quickly. Occasions have occurred when, during a survey for a sunken target, navigation systems failed just prior to "finding" the target, and since the survey team did not know where they were, the area had to be resurveyed. A constantly updated manual or computerized log denoting position is therefore important. When a target is found, and a navigation system becomes inoperative, you may couple the last known position with dead reckoning navigation to enable you to reacquire the previously found target and obtain a more accurate fix.

Prior to the early 1970's, navigational displays were often placed close to the sonar operator and he either kept a navigation log based on time, or wrote the navigational information directly on the sonar record. With today's digital sonar systems, the recorders accept an output from a large variety of navigation systems including microwave, loran and GPS. Using a navigational interface, position data is logged directly onto the sonar record hard copy and taped data without the operator's involvement.

Although some sonar operators still prefer to annotate a hard copy record manually or keep a time based navigation log, the use of a navigational interface allows the operator to analyze the side scan data without concern for navigational annotation.

OPTIONAL EQUIPMENT

Sonar systems are available with a number of optional modules. One of these, a *data storage unit*, is an important option, particularly when constructing sonar mosaics. Early sonar systems could only supply analog data for storage and were limited in capability. Newer systems have the capability of storing digital or analog data, giving the user the advantage of storing the full dynamic range of all the sonar data in raw or processed form. Other data that is now stored includes auxiliary information such as navigation, time, date, line and fix numbers, range, speed, and annotations along with various system operating parameters. Modern data recorders also have a very high dynamic range capability which gives the surveyor many options in post processing sonar data.

Data storage units are available in the form of either magnetic tape, or computer controlled magnetic and optical storage medium. They store hundreds of megabytes which can represent many hours of sonar data depending upon the range scale (scans per second) used when gathering the data.

Stored data is used for activities such as duplicating original hard copy data for the construction of mosaics, reports and presentations. Records can be regenerated from stored data to correct graphic presentations for distortions due to navigational errors, layback, and towfish position and improve image resolution through the use of image processing systems. Future storage media will provide faster and higher capacity modules for storing large quantities of sonar data. These modules will always be important for data processing after at-sea operations.

Color display of sonar data is possible through another optional module called a *video display*. These video displays use color to add another dimension to assist the operator the interpretation of side scan data. This is especially true for target identification operations such as minehunting or shipwreck search where the increased dynamic range of a video monitor shows subtle variations not evident on a hard copy printout. In both real time and post processing operations, video displays are helpful for improving target detection by allowing the operator to choose the optimum color pattern to accentuate selected targets.

The video display is often used with a *target logging system* that has the ability to store selected targets and then recall them later on the video display for comparison testing or post-processing analysis. Other salient features of video displays include target expansion, selectable color pallets, menu driven software, waterfall displays, and target expansion.

Although it has always been possible to run a number of marine instruments simultaneously during offshore operations, some instruments will interfere with the sonar data. Profilers, sparkers and other acoustic instruments can be run with side scan sonar but they cause patterns to be printed over the sonar data (see Record Interpretation, Chapter 8).

The magnetometer, an instrument commonly used in geophysical, geological and target search operations, is usually too sensitive to be used simultaneously with other instrumentation. For some survey operations a magnetometer can provide important information on seabed magnetics or buried targets but in the past, magnetic surveys were usually performed in a separate operation from the sonar work. Recently, a dual side scan/magnetometer system was developed that incorporates both instruments in one package. Careful integration has resulted in an instrument that supplies accurate information from both instruments. Called the Mag-Scan™ (EG&G), the design furnishes the surveyor with both acoustic and magnetic data from one survey. The instrument uses the third data channel of the sonar readout to display the magnetic data. It also sends the data to the data storage module if one is being used. Record 38 is an example record from the new system.

The development of the Mag-Scan™ is an example of advances in the marine instrumentation industry providing options, accessories and modules that will help the surveyor accomplish his task in less time and at lower costs.

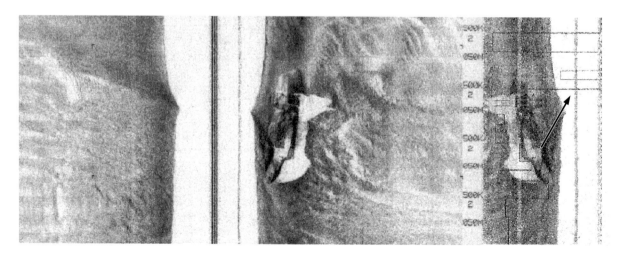

Record 38: A survey of a sunken wreck is shown above. The wreck is rapidly becoming buried in sand. As the site becomes covered, the value of the magnetic data increases. This combined sonar/ magnetometer system, displays the magnetic values in the third sonar data channel (arrow).

Chapter 7:
At Sea Operations

A sonar operation is dependent on a complex variety of information, equipment, and personnel. Sonar, once mobilized, works only as well as the operator's skill and the vessel and crew supporting it.

Weather and sea conditions are also key. If the survey area is exposed, operating during the most favorable season and "weather window" is recommended. If this is not possible than one must choose the most optimum time available and standby for a fair weather window. The more exposed the operating area, the more critically this relates to vessel size, time commitment, and funding. For example, a nearshore operation on a 20 meter vessel in the Northwest Atlantic Ocean during the summer months may provide a minimum of one out of three operational days. In the winter, the operation will probably be limited to one day in seven.

Time becomes even more critical when the survey is in exposed waters that are deep, contain unpredictable and significant currents, and when the area to be searched is large. Adequate planning is essential, and includes reviewing oceanographic and meteorologic conditions of the area and researching data that may better define the survey area. Other important items to be considered are: sonar frequency required, bathymetry within the operational area, surface and underwater positioning requirements, time allotted for the survey task, and ancillary equipment needed such as navigational, video and data recording interfaces.

SPARES

Operational planning requires careful consideration of the equipment that will be deployed once on site. The duration of the operation and the distance of the search area from a cargo airport, or from a facility that has spare parts will indicate what spare parts for the required search operation should be available on site. Spare parts may even include a spare towfish, cable and full electronics compliment. On the other hand, during an operation close to shore and support facilities, few spares may be required. Electrical spare parts for digital side scan usually are not more detailed than PC board spares since these new systems are difficult to troubleshoot to the component level without specialized test equipment.

In shallow water search operations, or those where there is a sufficient budget, a second towcable is often an important spare part. Towcables fail for a myriad of reasons, from a failed underwater connector to Z-kinking failure at either end of the cable length. Whatever the reason for cable failure in a multiconductor cable, it can cause the loss of one or more channels of data required for quality sonar imagery. If a spare cable is immediately deployed while the primary cable is repaired, hours of survey time are salvaged. Frequently this "down time" occurs during optimum weather conditions. During deep sonar operations or those using fiberoptic or single conductor cable, a spare cable will save days of "down time" while the failed component is being spliced.

The sonar towcable is a very important part of the system. The towcable is also vulnerable because it is exposed to a large portion of the water column with all the inherent dangers of mooring lines, ghost fishing gear and fishbite. The cable terminations, an underwater connector on the "wet" end and a multi-pin MS type connector on the "dry" end are the most susceptible to damage and care must be taken during transport, storage and use. Spare connectors and terminating materials must be on hand for any serious sonar task. The proper choice of cable must be made in the planning stage. Armored cable is a nuisance to transport and handle, if it is not required, and lightweight cable cannot be brought into deeper water due to its low density and inability to withstand the tension of depressor application.

The wet connecter on the towfish end of the cable can even fail from general wear and tear. If the connector fails during an operation it requires replacement with either a spare cable or a retermination before work continues. When connector failure occurs, the data may be affected in a number of ways.

Often a failing cable will provide an indication of imminent failure before a complete loss of continuity. In other cases the assembly may fail completely in a matter of minutes. Typically, in the case of multiconductor cables, failures either occur with the power/ground conductors or it occurs with the returning signal conductors. Records 39 and 40 are examples of connector failure. In Record 39, one channel's signal leads failed affecting only that side of the sonar. In Record 40 the power or ground conductor failed and both channels were affected. Although the survey operation could continue after only one channel failed, the second example required immediate cable or connector replacement. In these cases, the problem was rectified by reterminating the towcable.

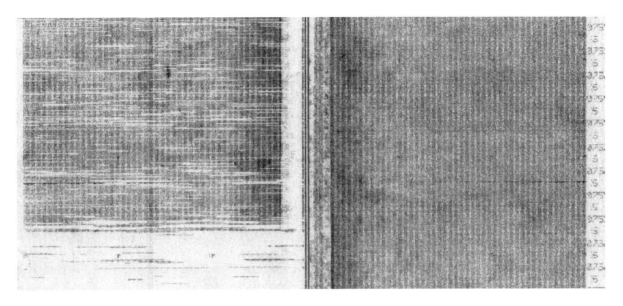

Record 39: The sonar cable is a fragile link between the towfish and the system recorder. Failure sometimes occurs with the wet connector on the towfish end of the cable. In the case above, the signal leads of the connector failed and only effected one channel of the sonar. The fault was intermittent but required retermination of the cable and connector assembly.

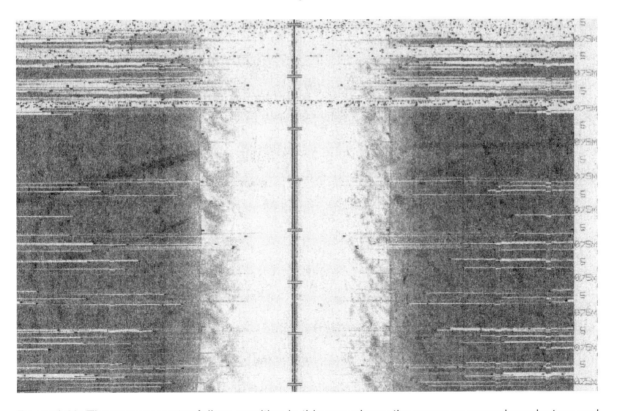

Record 40: The wet connector failure resulting in this record was the power or ground conductors and resulted in simultaneous failure of both channels.

TESTING

One of the most important parts of mobilizing portable sonar operations is the testing of all non-redundant equipment on shore. This procedure is important because, when offshore, an operational team often has limited access to necessities. Diagnoses and repairs, if necessary, would have to be accomplished in unfavorable conditions. Being at sea means being without replacements.

Once the standard operational preparations have been made in the lab according to the operations manual, the sonar is transported to a ship during which time it is often handled by a number of people. Once on the ship, the system should be thoroughly tested before the vessel leaves the dock. This includes mounting the recorder where it will be used during the main operation and assembling all the other system components at the locations where they will be used. Cables should be run and secured, ancillary equipment tied down and the tow cable set for deployment. The cable should be connected both mechanically and electrically to the towfish.

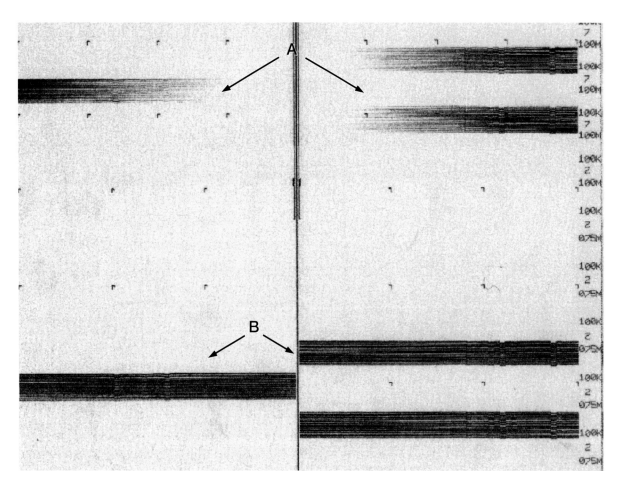

Record 41: Rubbing side scan transducers before deployment will generate noise that is detected by the sonar system. At position (A) the TVG circuitry is on and the noise appears highest at the outer edges of the record. At (B) the same test is performed with the TVG turned off and the gains driven high. The rub test is helpful in checking that the system components (towfish, cable and recorder) have electrical continuity.

One of the first tests, which determines if the power and signal connections are intact through the entire system, is call the *rub-test*. The operator should turn the system on, manually increase chart speed to 4 or 5 knots, set the range to 150 meters, set the gain to maximum and turn the towfish trigger switch on. He should then have an assistant firmly rub one transducer back and forth with his hand until he sees a dark line on that channel of the chart. The operator must confirm which channel is being rubbed. The assistant, on command from the operator, then repeats the process for the other transducer, and finally, again for the original side. This test confirms the proper operation of the TVG circuitry. The lines on the chart paper should resemble those in the upper part of Record 41 with most of the noise appearing at the outer ranges of the record. The same test should be done with the towfish trigger turned off. Turning the trigger off drives the TVG circuitry to maximum gain and the chart printout should resemble the lower portion of Record 41. It does not matter whether the rub-test is done "starboard-port-starboard" or the reverse. But it is important that the operator know which side is done once and which is done twice. Repaired towcables are sometimes wired with the channels reversed. During operations using an improperly wired cable, objects that are to the starboard side of the survey vessel will appear on the port channel of the side scan sonar. More than one sonar operator has been perplexed when he instructed the bridge to turn the vessel about to investigate a target to port and find that it is not there. The rub-test not only confirms the electrical connections are intact, but also that the cable is wired properly and that the TVG circuitry is operating. Most

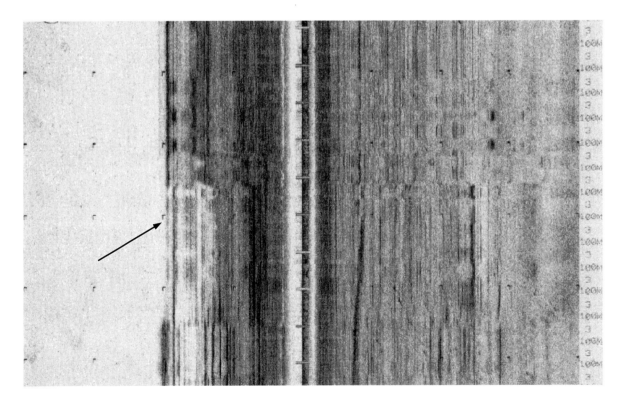

Record 42: Checking the water tight integrity of the overboard equipment is another test that can be performed before the vessel leaves the pier. Although the images generated are smeared, aiming one side of the towfish towards a shoreline (arrow) or other targets helps confirm system operation.

new side scan systems have "self-tests" that should be performed in order to assure that other components of the system are operating properly. The operators manual usually outlines these procedures and how they should be executed.

To test the integrity of the electronics housing seals and the underwater connector on the towfish, the fish should be lowered at the side of the dock and records printed. Since the towfish will not be moving over the seabed, the records will only reflect the echoes of the immediate surroundings. The record generated will appear smeared as in Record 42 where the port transducer was directed at a shoreline 50 meters away. This test provides the operator with confirmation that major mechanical and electrical subsystems are functioning. Since towcable and connectors are the most common failure points in a system and repairs are performed far more easily on shore, if any faults are detected they can be corrected before the vessel is at sea.

SYSTEM DEPLOYMENT

The process of deploying a sonar system very quickly becomes second nature to those who frequently go to sea with remote sensing equipment. However, there are a number of steps that should be taken to avoid damage or loss to the towfish during deployment. The system should have a brief check, similar to the one at dockside. The operator should personally double check all shackles and fittings that attach the towfish to the ship. He should also visually check that all components are secure and rigged properly. A second "rub test" is recommended if the system has experienced any shock or vibration since leaving dockside. When these checks are complete, and the survey is to commence, the recorder should be turned on. The chart speed should be set at 4-6 knots to reduce any delay between the printing on the paper and the time when the image is visible to the operator. The range should be set to two or three times the water depth. This is important because the operator needs to confirm the acquisition of the first bottom return after the fish is in the water and before lowering begins in order to prevent collisions of the towfish with the bottom. The recorder may log both the bottom return and the surface return in rough seas when the first surface return is usually strong. This can become confusing. But when the towfish is lowered the two signals will be differentiated as the usual harder sea bottom return becomes more distinct than the surface return. With the survey vessel at low speed and holding a steady course into the sea, the fish should be eased into the water so that it comes around into the tow path behind the survey vessel. Care should be taken to pay out enough cable during launch to keep the towfish away from the ship's propellers. Once the operator has confirmed acquisition of the first bottom return, the range setting on the recorder is lowered if desired. However, the operator should not set the range to a figure less than the water depth during lowering. This range decrease should take place gradually as the fish nears the bottom in order for him to keep visual track of the fish height. The towfish is lowered until its height is approximately 10-20% of the range setting that will be used during the survey. The survey vessel then comes up to survey speed and the fish height fine tuned by retrieving or paying out cable. After the fish is at depth, and the ranges are set, the chart speed is changed to represent the true speed of the vessel and the system is tuned.

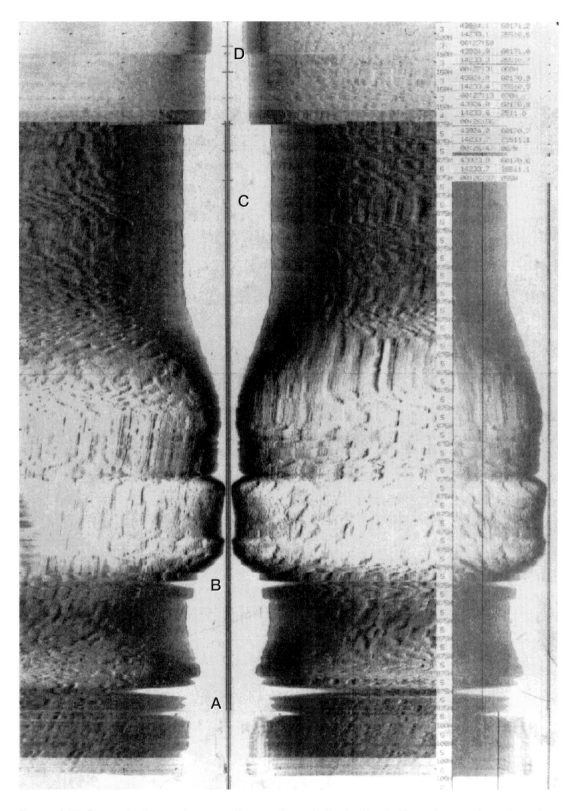

Record 43: Some shallow water operations or those in fluctuating bathymetry, use in-water cable lengths that are shorter than the water depth. With the survey vessel stopped, the towfish is lowered (A). Lowering slows as the towfish nears the bottom (B). When the survey vessel gets underway, the fish comes to altitude (C) and the recorder is set to the proper ranges (D) and tuned.

For operations in restricted waterways where the survey vessel may be forced to make sharp turns with little warning, the in-water-cable length should be less than the operating water depth in order to prevent seafloor collision. This may mean that the fish is towed higher in the water column than the prescribed 10-20% of the range settings. Deploying the system in this case would mean bringing the survey vessel to a dead stop in the water. The cable is then payed out until the fish is just above the bottom. The operator can closely monitor this process by examining the sonar record being generated.

In Record 43, the deployment process is shown. In 24 meters of water the sonar is set to 100 meters per side and the fish is lowered. The operator sees the first bottom return and then shifts the recorder to 75 meters per side. Lowering commences as seen by the jagged first bottom return (A). This is caused by the nose-down pitch of the towfish as it is lowered. As the fish nears the bottom (B), lowering is halted and fine tuned. With the proper amount of cable out, the survey vessel comes up to course and speed and the cable drag in the water brings the fish up to height. The correct speed is fed into the system and annotation begins (C). The recorder is set to the ranges to be used on the survey and the gain setting is changed to provide a good seabed image (D).

TUNING

Tuning a modern side scan sonar to generate quality records is much less complex than it was only ten years ago. Sonars have also become "smarter" and are able to produce higher quality imagery. As a result, it still requires a certain skill level to maximize the quality of today's side scan systems. On newer side scan sonars, tuning is a reasonably straightforward process using printer gain controls.

To test system tuning, the surveyor should use an area with moderate depth (15-35 meters) and a clear bottom with some targets such as sand waves, rock outcroppings or a shipwreck. The fish should be at a reasonable height for the water depth and range settings used. For general survey, the sonar records should show an even, light gray, edge to edge, and on video systems, an even color hue. Targets should show up as distinct acoustic reflectors and shadows should remain anechoic. Too dark a record will mask target detail and too light a record prevents shadow detail from being recognized. Mixed bottom conditions may provide an image with darker and lighter areas. When encountering varying bottom conditions, the operator may find it necessary to change the gain settings. This is done to match the dynamic range of the recorder to the dynamic range of the sonar returns and aid the maintenance of an easily interpreted sonar record. Record 44 shows a evenly textured seabed ideal for system, operator or vessel captain training. Record 45 shows a flat seabed using different gain settings from the lowest to the highest. Most often the proper gain setting for any particular record depends on the eye of the operator. When scanning targets and achieving accurate positioning of components within a complex target such as a shipwreck, the operator should try a variety of gain settings. Often, operators make the mistake of setting gain controls too high. It is easy to err in the other direction as well, but as a rule of thumb, a lighter record will reveal more detail of a complex target than a darker one. Gain controls that are too high mask the subtle details of components within a target. By utilizing a combination of contrast and gain controls, the user

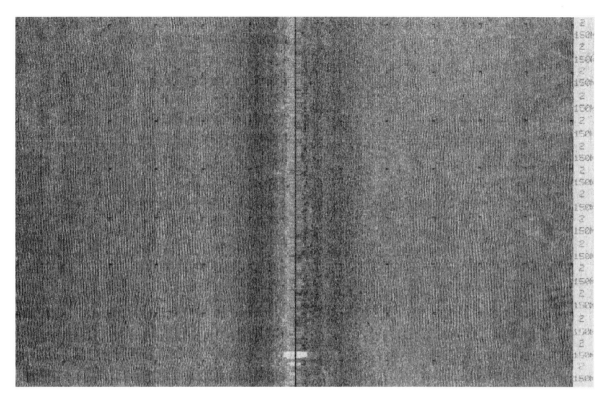

Record 44: The ideal seabed for systems testing or operator training consists of even, smooth, flat bottom with some targets. Seabed irregularities such as sediment ripples as in the record above helps the operator become familiar with the various settings available on the system.

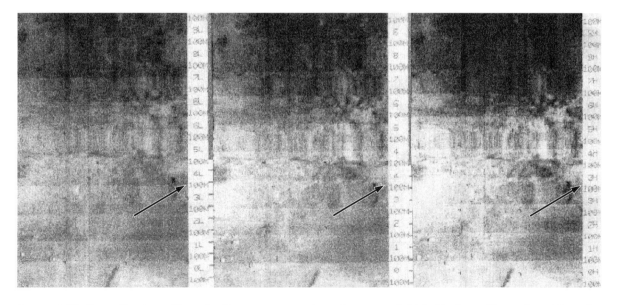

Record 45: Overall gain settings for the sonar record can be seen eight 3 dB steps. The gain settings for this record can be seen in the system status line (arrow). This record is actually a composite of three records to illustrate the advantages of using contrast and gain settings together (see Optional Settings).

can control the image definition of the record more effectively (see contrast under *Optional Settings* section). Record 45 is a composite of three images. In this composite the three records show the effect of low, normal, and high contrast.

Records are made in the slant range corrected or uncorrected mode. Both have their applications in different sonar operations. The use of the corrected mode in generating sonar records, as is seen in Records 21 and 23, both removes the water column and provides a lateral accuracy in the seabed image. For survey work requiring mosaic construction or applications where lateral distortion should be removed from the data, corrected records should be generated during the survey. For search operations where the target is small or very deteriorated, uncorrected data is more accurately interpreted.

As described in Chapter 5, paper speed should be set for most operations to accurately reflect the speed of the towfish over the bottom. This is done either manually, or, automatically if a navigational interface is used. The accuracy of automatic speed correction through the use of navigational instruments depends upon which instrument is used to generate the speed information. Some instruments, such as Loran, have significant delays in generating accurate speed figures through turns and other course changes. In such cases, entering the speed manually is usually sufficient for gaining accurate records.

OPERATIONAL RANGE SELECTIONS

The ranges used in sonar surveys are very important and should be selected with forethought during the operational planning stage. The parameters of the operation will indicate to the careful surveyor what ranges should be used. Long sonar ranges, by definition, utilize fewer outgoing pulses per second than shorter range settings. This limits system detectability of objects on the seafloor as well as operator recognition of such targets. Further, the longer range settings scan a larger area of the seafloor and, since the display is a fixed size, compress the resultant images. This results in targets being smaller on the final display and thus they may be more difficult to recognize. For 100 Khz operation, long ranges, up to 600 m per side are used if the requirements of the survey allow. Operations using 500 kHz very rarely use more than 150 meters per side and more typically use less than 100 meters per side due to the shorter range capability of the higher frequencies. (See *Target Search Operations* later in this chapter.)

LANE SPACING AND OVERLAP

For virtually all side scan sonar operations the survey vessel must follow a series of imaginary tracks or *lanes* on the earth's surface. These lanes should be parallel and follow a specific repeatable direction. They are most often generated on a survey plan based on a navigation system such as GPS, loran or microwave positioning. The *lane spacing* (distance between two successive tracks) is determined by the requirements of the survey.

During target search operations, the sonar is operated on a range determined by the survey requirements or the target size and anticipated condition. The survey vessel also has some error in following the track. The lane spacing must be less than the *swath width* of the sonar. This will allow for *range overlap* and assure total

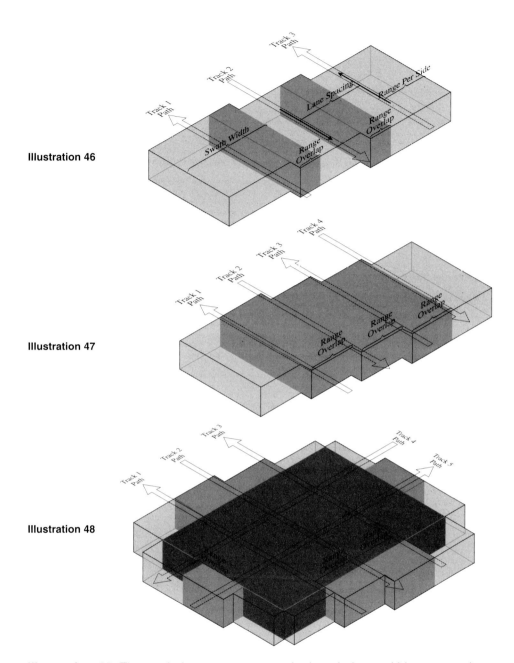

Illustration 46

Illustration 47

Illustration 48

Illustration 46: The track that a survey vessel takes during multi-lane operations must allow sonar coverage to overlap the previous lane. This forgives vessel track variations and increases target recognizability in the event that a target is encountered at the outer ranges. Illustration 46 shows a lane spacing of seventy five percent of the swath width which, on all but the first and last lanes, gives a seabed coverage of one hundred fifty percent.

Illustration 47: This more conservative methodology uses a lane spacing of fifty percent of the swath width which provides an inner lane coverage of two hundred percent, and is often referred to as 100% overlap.

Illustration 48: A very conservative methodology covers an area using track lines at 90° to one another. The first swaths cover the entire search area using a lane spacing of fifty percent of the swath width as in illustration 47. The second swaths use the same lane spacing but are run at a right angle to the first.

coverage of the area being surveyed as well as compensate for the inherent loss of transverse resolution at the outer ranges. One determination that must be made prior to commencing the survey is the exact amount of range overlap required.

Seabed insonification during search or survey using side scan sonar is described in terms of lane spacing, range overlap or percent coverage. Three illustrations should be examined to clarify these descriptions (Illustrations 46, 47 and 48). The tracks and lanes depicted in these illustrations are abbreviated in length and would be much longer in an actual survey. For large targets in smooth seafloor and operating conditions, a lane spacing distance equaling 75% of the total swath width is often sufficient to overcome the vessel tracking error. This also assumes the use of a reliable, accurate navigation system and experienced vessel pilot. If one assumes that one single pass of the sonar provides 100% coverage of a given area, and the vessel follows track exactly, a lane spacing of 75% of the swath width provides 150% coverage of the seabed (Illustration 46). Although vessel track and navigational errors rarely allow exact tracking precision, this lane spacing is 75% of the swath, which is equivalent to 25% overlap or 150% coverage. In other words, 50% of each area covered, is covered again by previous and successive lanes. In the proper conditions, where time is at a minimum, a careful operator can perform effective target search or seabed survey with an overlap of 25%.

One of the most common lane spacings used for target search is a lane spacing of 50% of the swath width, having 50% overlap or 200% coverage on all lanes inside the two outside lanes. Figure 47 shows the coverage on such a survey. This method of surveying brings the outer ranges of the channel covering new seabed into the near ranges on successive passes and also provides a large margin of error in vessel tracking.

An extremely conservative area coverage uses two different tracklines. The area is essentially surveyed twice. The first area coverage uses the 200% coverage described in the previous paragraph. The second coverage of the area repeats this range overlap, using tracklines at 90° to those used in the first coverage, shown in Illustration 48. This provides for 100% overlap and 400% coverage of an area. It is particularly advantageous to insonify an area using tracklines at 90° to one another because many targets have one axis longer than the other. If the target happened to be difficult to recognize due to its perspective when insonified, with the new tracklines, it could then present a better profile for imaging (see Records 30 and 31).

OPTIONAL SETTINGS

When generating corrected sonar records, the height of the towfish, the angle of the sonar beam intersecting the seabed and the makeup of the seabed affects the strength of the signal return in different parts of each scan and the overall evenness of the record. The grazing angle control assists the operator in getting an even record by providing different seafloor models to the processing electronics during image generation. Record 49 shows the effects of different grazing angle control settings on one specific seafloor environment. The operator should experiment with the settings to determine which gives the most balanced overall record.

Even though an operation requires corrected sonar records, it is advisable to deploy the sonar system in the uncorrected mode. This allows the operator to examine closely, the water column and fish height information during fish lowering. Record 50 shows the record as the operator generates and fine tunes a corrected record. Since corrected records usually require slightly higher gain settings than uncorrected, the gain is turned up by one setting (A). The altitude is switched from manual to automatic, feeding true altitude information (taken from one sonar channel) into the system thus removing the water column from the record and automatically correcting for slant range errors (B). The operator then sets the grazing angle to provide an even seabed image (C).

Along with the grazing angle settings, the side scan recorder also provides settings for contrast enhancement and inverse printing of the sonar images. The contrast enhancement is useful in underwater environments where the overall bottom conditions are of very low or high contrast. Although most operations are run with the contrast set to Normal, if bottom features do not display adequate contrast, a setting of High is selected to make subtle features of the bottom terrain more recognizable. Conversely the Low setting will lower the contrast of sharply mixed bottom characteristics, providing a more even record. Contrast is used in combination with gain controls to allow greater image control (see record 45). Record 82 of fishing trawler door marks on a smooth seabed was generated on the low, normal and high contrast settings. In this low contrast environment, the high contrast setting helped delineate the bottom features.

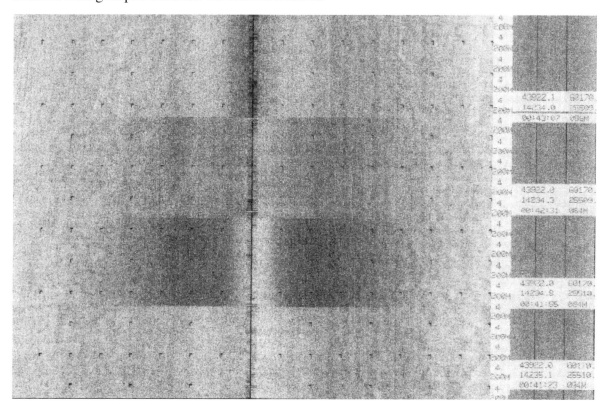

Record 49: Grazing angle changes will provide the operator with an even record when operating in the corrected note. Fluctuations in fish height and seabed reflectivity often require adjustment of the grazing angle control. The record above shows four different settings with a constant seabed-towfish geometry.

Inverse printing of the sonar data displays strong sonar returns as white and the weaker ones darker shades of gray to black in the shadow zones. Sonar operators who have become accustomed to hard returns displayed as dark and shadow zones as white typically do not use the inverse mode. Some operators however, find the inverse record easier to interpret. This feature does not effect the data being recorded and targets will be as noticeable on an inverse record as on a normal record. Records 51 and 52 show the effects of a normal and an inverse sonar record. The seabed in both records is the same, consisting of flat smooth sand with large boulders. The records are very similar except that one is a "negative" of the other.

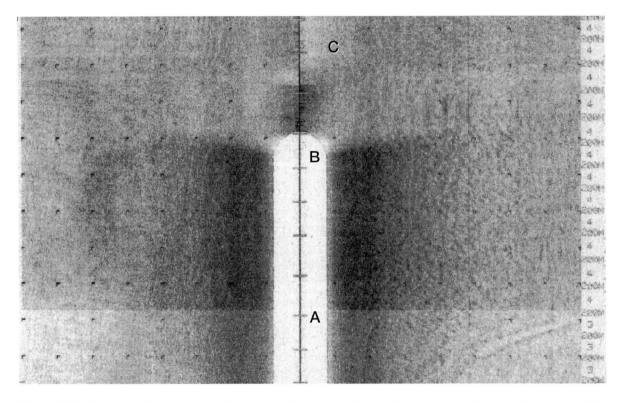

Record 50: Sonar towfish are often deployed with the recorder set to manual so the operator can readily determine the towfish height during lowering. Once deployment is complete corrected records can be generated. Typically the gain is set higher for corrected records (A), the altitude is set to automatic (B) and the system begins to track the bottom, removing the water column and generating slant range corrected records. The grazing angle is then set to match the seafloor and give an even display of the bottom (C).

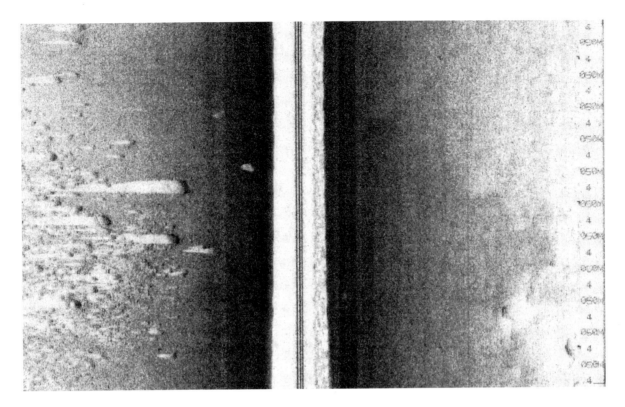

Record 51: Shown in the standard mode this side scan record was made over a plain sand seabed with large boulders on the port channel. In this mode targets appear dark with white shadows behind them.

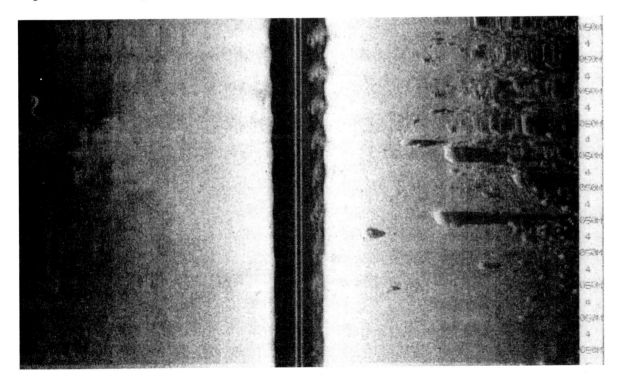

Record 52: Records in the inverse mode contain the opposite shades from record 51 with targets displayed as light images casting dark shadows. This record was made over the same seabed as record 51.

TOWFISH STABILITY

Side scan sonar is capable of generating quality records in a variety of environmental conditions. For operations other than detail mapping, (see chapter), the system is tolerant of a number of factors that affect *towfish stability*. In general, however, towfish stability is a factor that should be considered for increasing the accuracy of the images generated. Similar to a still camera with a slow shutter speed, a sonar generates blurred, distorted and inaccurate images from a severely unstable towfish.

Many environmental factors contribute to towfish instability. These are most often expressed in the form of *heave* and *pitch* (see Illustration 52A). Since the towfish is directly connected to the survey vessel, the movement of the vessel on the surface sometimes dictates the type of movement the towfish will experience. Waves on the surface of the ocean cause pitch and heave of the survey vessel and translate to heave in the towfish. Heave affects the sonar imagery by causing a rhythmic increase and decrease in altitude of the towfish and also a degradation of the image of seafloor targets. In severe cases, when the towfish altitude changes swiftly and significantly, the range to bottom target changes as well. As a result images appear broken and disjointed.

Although most well-designed towfish allow for a gravity stabilized, ballasted construction, the very fact that they are designed to be towed (horizontal movement), and have stabilizing fin assemblies, prevent them from providing quality imagery during severe heave (vertical movement). When the towfish heaves because of the influence of surface instability, some pitch is also induced. This is often evidenced by lighter streaks on the sonar record where the seabed was not properly insonified. The effects of heave and pitch are often seen together. Record 53 shows heave in a towfish due to wind driven sea surface roughness. In this record, pitch is more noticeable than the heave as evidenced by the minimal change in fish altitude shown by the first bottom return. Record 54 also shows the effect of pitch and heave in a corrected record. Severe heave, as experienced during the generation of Record 55, significantly degrades a target image. The fluctuations in fish height are seen in the first bottom return. Pitch, as a by product of heave, is evident in the white streaks in the record. The target shown here is the same shipwreck shown in Record 143.

Roll is another form of fish instability with an effect on the record similar to the effects of heave and pitch. High frequency roll will also result in white streaks on the record and it may be difficult for the operator to determine exactly which instability is occurring. Pure roll by itself is not frequently seen in a sonar record since the conditions that would cause excessive roll causes other instabilities as well. The towfish will roll during turns, however, aiming one transducer higher than the other. The effect of roll can be seen in Record 115 when the towfish banked during a turn.

Even mild heave, caused by long wavelength fluctuations in fish height, impart distortions in a sonar record. Ocean swells in coastal zone waters create heave that may not be readily evident in the first bottom return but will distort target images. Record 56 is of a flat soft seabed and a submarine. The snake-like curves in the target are a result of low frequency fish height changes caused by heave.

Yaw is a rare towfish instability that does not normally occur. Yaw is a side-to-side movement of the fish about a central axis. This is most often caused by a damaged or lost tail fin assembly. This instability causes the sonar to scan one area for a longer period than it should and then rapidly move ahead scanning the area passed for less time than it should. Effectively the towfish experiences the same conditions as if it were making many small turns repeatedly. Yaw-like movement in wide swaths can occur using long cable lengths and depressors. This movement is referred to as *kiting* and is a result of damaged or improperly attached depressors.

Pitch sometimes occurs without the normal effects of heave but it is rare. In Record 58, the effect of pitch is seen as evidenced by the fine white lines across the sonar record. The pitch here was induced by a faulty depressor.

Currents also affect the sonar towfish in a number of ways. Small eddy currents that occur around bottom structures temporarily destabilize the fish and create distorted records. These eddy currents are often seen on the surface of the water as evidenced by upwellings. Sometimes the effect of eddy currents can be ignored, particularly when generating images of large seabed area, but for detailed mapping these eddies are problematic in generating accurate imagery.

More consistent, large scale currents cause problems for the surveyor. Such currents change the heading of the towfish so that it has a different heading than it's track over the bottom. This is particularly complicated in deep operations where slow tow speeds are used. Since the surface vessel may have to "crab" into a surface current it's heading also may not match the survey trackline. This has the overall effect of narrowing the sonar coverage and also results in imprecise seafloor target positions. The longer the range setting on the sonar, the more serious these problems become. The only solution for this poor *path-tracking* of the towfish is to examine the currents in the survey area and modifying the survey lanes to minimize their effect. If target positions seem inconsistent, insonifying them with or against the current will improve the intended towfish path-track and provide more accurate position information.

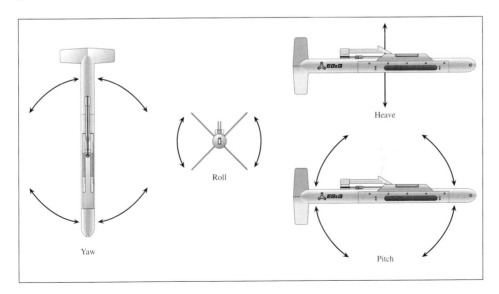

Illustration 52A: Towfish instabilities take a variety of forms as shown above.

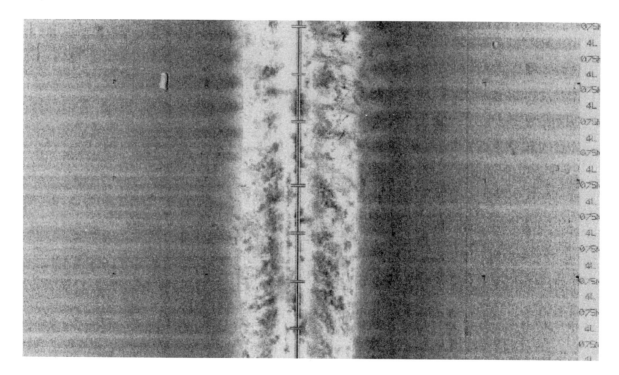

Record 53: The most frequently encountered towfish instability is heave caused by pitch or heave in the towing vessel. This instability is apparent in the sonar record as a fluctuating first bottom return and image curvature of straight targets. Very often heave in the towfish is accompanied by other instabilities such as pitch and yaw. These latter motions classically cause lighter streaks on the sonar record.

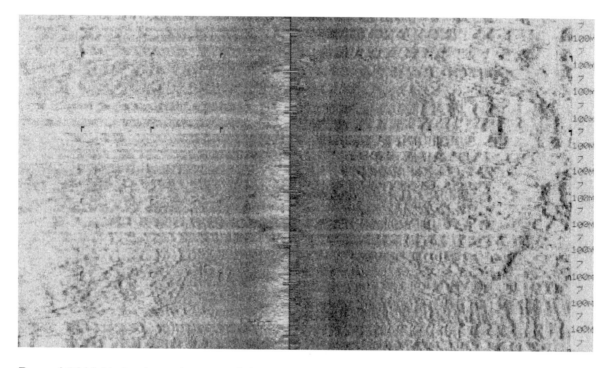

Record 54:Light streaks and a general degradation of the displayed images are the most noticeable indication of a towfish experiencing severe instability when operating in the corrected mode. This is typically caused by rough sea surface conditions. Although the effect is not pronounced in the above record, the image of small targets in heave can be degraded beyond recognition.

Record 55: Even large target images can be significantly degraded from survey vessel heave and the resulting towfish instability. This target is the same shipwreck as that imaged in Record 143.

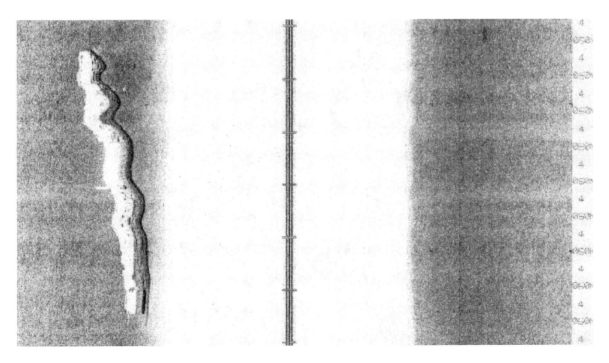

Record 56: Low frequency heave, such as that from ocean swells, may not induce significant pitch in the towfish assembly. This eliminates telltale white streaks across the record while the heave component will still degrade target images. The image of this sunken submarine is curved due to low frequency heave in the towfish.

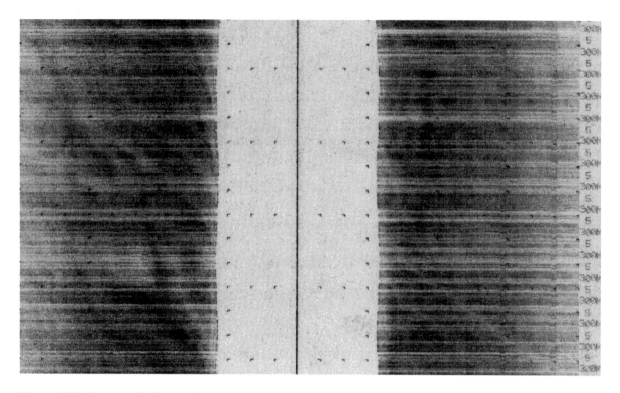

Record 58: Severe pitch sometimes occurs without heave in the towfish but it is rare. The white streaks in the above record are from pitch in the towfish induced by a damaged depressor.

FISH HEIGHT AND POSITION

The quality of the sonar data is often a function of the height of the towfish above the bottom, or bottom targets, during a survey or target imaging. In general, with standard sonar configurations, surveys are performed with the towfish positioned a distance above the bottom approximately equivalent to between 10% and 20% of the range setting of the sonar. This will depend upon the tilt of the transducers down from horizontal. Most side scan systems are designed with the transducers tilted down from horizontal between 10-20 degrees. More acoustic energy is then dispensed toward the seafloor when the fish is flown at normal working altitudes. The tilt to the transducers is often adjustable for working at higher and lower altitudes.

The optimum height of the fish is a function of range, because at the outer edges of the record, the angle of incidence of the outgoing acoustic pulse on the seafloor is one of the factors that determines the amount of *backscatter* and returning echoes to the towfish. If the transducer array is towed high off the seafloor, shadowing will be lessened and target recognition may be reduced. If towed too low, the reflectivity at outer ranges will be reduced limiting the effective range of the system. Illustration 59 shows the general effect on reflectivity for different intersecting angles of sound with the seafloor. In the illustration, different seafloor faces are shown but this effect is also seen by changing the fish height.

There are conditions which necessitate towing a fish higher than recommended, such as in greatly fluctuating bathymetry, or lower than recommended, such as in very shallow water or when enhanced shadowing is desired. The operator should be aware of the towfish height effect on the resulting data.The fish is usually flown higher in deep water when using long lengths of cable and wide swath-width sonar systems because of the lack of towfish rapid response to winch activity. In many of these systems the outgoing pulses are directed downwards slightly more than in conventional systems. For towing short lengths of side scan sonar cable a good rule of thumb is that the operation will require a cable length equivalent to three times the water depth to position the towfish at the appropriate height.

In the past, determining the length of cable needed for a particular towing configuration, the amount of depressive force required to position the towfish and the amount of tension at the ship, required some educated estimates coupled with trail and error deployments. Towing configurations are quite straightforward during operations using relatively short cable lengths (ie. <50m), however as the amount of in-water cable increases, currents and other environmental factors increase the complexity of predicting towcable behavior. Small, high speed, personal computers provide a means to use a number of tow cable prediction software packages that have been designed to assist the surveyor in proper planning. Cable prediction software packages accurately predict the behavior of side scan towcables in different current situations even with a variety of depressors attached.

Keeping the towfish from striking the bottom is very crucial. As a result it is very important for the sonar operator to be in constant communication with the bridge and/or helmsman to prevent an accident. Although the sonar operator would most often be communicating with the winch or cable tender, he must always have an open line to the bridge as well.

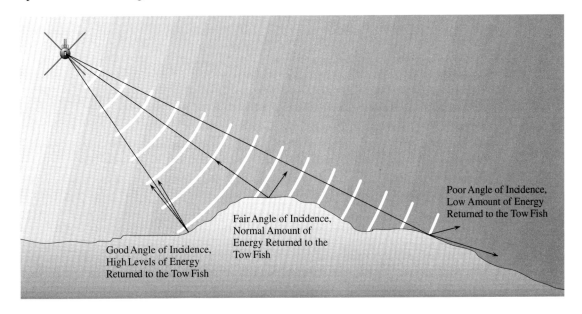

Poor Angle of Incidence,
Low Amount of Energy
Returned to the Tow Fish

Fair Angle of Incidence,
Normal Amount of
Energy Returned to the
Tow Fish

Good Angle of Incidence,
High Levels of Energy
Returned to the Tow Fish

Illustration 59: The reflectivity or backscattering potential of many underwater surfaces is a function of the sonar beam's angle of incidence as it encounters that surface. When the sound pulse is normal to a surface, far more energy returns to the towfish than when the beam strikes a surface at a large angle. This angle of incidence along with the surface roughness are the primary reasons for dark and light areas on the sonar record. The various intensities of these shades help with accurate record interpretation.

In most bathymetry operations, acoustic data acquisition is performed using a transducer mounted to the survey vessel. Accurate data acquisition is easy to achieve because measuring the distance from the navigational antenna to the transducer is a simple task. In side scan sonar operations, with the transducer assembly mounted on a towfish that is astern of the survey vessel, knowing its exact position may not be as straightforward a task. For shallow operations the towfish *layback*, or range behind the vessel, is estimated fairly accurately. However, as the amount of in-water cable increases, this becomes more and more complex. In areas of low current and consistent underwater conditions, and if the survey vessel has high accuracy surface positioning, the layback is estimated using bottom target detection. To accomplish this, the surveyor picks a distinct underwater target as it is detected by the sonar and notes the survey vessel's location. Then, making a pass in the opposite direction, with the same amount of cable out and the towfish kept at the same height off the bottom, he notes the vessel position when the target is again detected. By dividing the distance between the two noted positions that surveyor estimates the layback of the fish. Depending on the accuracy required, this method is reasonably effective using up to 500 meters of cable. However, for areas where inconsistent underwater currents or cross currents exist that may effect the fish's tracking behind the survey vessel, an acoustic positioning system is necessary to get accurate range and bearing to the towfish and thereby to the target.

TARGET SEARCH OPERATIONS

During target search, the major function of the sonar interpreter is to not only distinguish man-made objects from natural ones but also to distinguish untargeted man-made objects from that which is the object of the search operation. The proper analysis of targets during search operations is time consuming, but saves valuable search and identification time. For search operations to locate objects of most sizes (> 3x3 meters), the frequency chosen is 100 kHz using ranges from 100 meters for small targets, to 600 meters for very large targets.

The ranges used are a function of several variables. The use of longer ranges during search operations decreases the time required for the search. However, since a given target may be more difficult to detect and recognize when using long ranges, the surveyor should avoid using ranges that would prevent him from finding the target. The size and condition of the target should be considered since larger and more intact targets can be detected and recognized using longer ranges. The topography should also be considered since a target will be more evident on a smooth seabed than a rough one. (see Records 64 and 65). Other factors to be considered during the operational phases of a search operation include the depth of the water, sea state and tow speed.

When a target is recognized on the sonar record, there needs to be a method of recording the position of the vessel, the direction of tow and the distance between the vessel and the towfish (fish layback). The first two parameters can be inscribed by pen or pencil directly on the sonar record, or for faster and more positive data recording, the navigational interface within the sonar recorder annotates the record. The layback of the cable does not change frequently during most shallow water search operations and this is recorded manually. Knowing the layback is important because the image generated on the sonar record is generated from the towfish which is almost never in the same position as the surface vessel. Later analysis of search operation records can uncover targets that require further investigation. The layback figure will assist the interpreter in determining the position of these targets.

During search operations for man-made targets, image repeatability is an important factor. As discussed in previous sections, sonar can exhibit unusual images caused by one or a number of acoustic phenomena. These images are often baffling to the operator, particularly when they seem to show "real" targets that are simply a result of non-man-made targets or the physics of sound in water. For this reason the operator needs to repeatedly scan suspect targets in order to properly assess their existence, makeup and true position. These baffling images could be caused by thermoclines, dense schools of fish, tow vessel turns or towfish instability. Usually an attempt to repeat these images will assist the operator in interpreting the sonar record. Schools of fish characteristically move and change configuration on the bottom. Thermoclines usually effect the record at outer ranges and vessel and towfish path changes "shift" them on successive passes. Target repeatability will allow the sonar operator not only to determine the existence and configuration of a target but also provide him with target position data that becomes more accurate with each pass.

Records 60, 61, 62, and 63 show a series of passes by a target located in coastal zone waters on a flat seabed. Record 60 shows the first contact (A) with the target while running on a 100 meter sonar range. The sonar operator was in constant communication with an operations manager who made decisions on the path, direction and placement of the survey vessel on each successive run. The survey vessel reversed course and ran a pass 15 meters closer to the target. The vessel wake from the original pass (B) then became visible on the sonar record. Note that the operator did not reduce the range of the sonar for this second pass but left it on 100 meters per side. Maintaining the same sonar range on the second pass is important to efficiently position the target. The second pass is often sufficient to get a close approximation of the target's actual position. Once this position is established the range is gradually dropped on successive passes.

Record 61 represents the third pass with the survey vessel passing even closer to the target. For this pass, enough position data was acquired to get the survey vessel very close to the target and the sonar operator dropped the range to 50 meters. On the fourth pass, shown in Record 62, the range was dropped further to get a more defined target image. Here, the target height was measured, and the next pass was designed to take the towfish directly over the target. Record 63 shows the target in the towpath of the vessel. It is seen that the target is cylindrical and measurements taken from the Records 62 and 63 demonstrate that the object has the approximate dimensions of a 10,000 gallon fuel storage tank.

Accurately interpreting sonar records during target search is complex and takes experience and patience. Overall, the sonar operator should be conservative in his interpretation and cautious in his target analysis. Even the best sonar operators are deceived by targets that appear exactly like the ones for which the search was initiated.

Keys to successful target search include looking for inconsistencies in the surrounding environment. For instance, some key indicators of target existence in the record might be: straight appearing objects within a boulder field, large clump-like objects within a seabed made up of smaller objects, or scours from a buried target in an otherwise featureless bottom. As divers attest, straight lines rarely occur in nature and their presence is the first key to man made debris on the bottom. In a difficult search area such as when searching a rocky area for a deteriorated target, keys to the presence of a target are difficult to recognize. This is particularly true when the operator has been studying sonar data for many hours in "challenging" terrain and becomes a victim of fatigue.

Record 64 was generated during a search for a steel shipwreck expected to be heavily deteriorated. The search area was in shallow water and the bottom was strewn with large boulders. Fish height was insufficient to run long range due to the shallow depth of the water. A range of 75 meters per side was the longest range expected to provide detectability at the outer portions of the record. The rocky terrain hides much of the shipwreck structure but a key to the presence of a man-made target is in the straight lines visible in the portion of the record marked A. The image from a consequent pass directly over the target previously seen to port

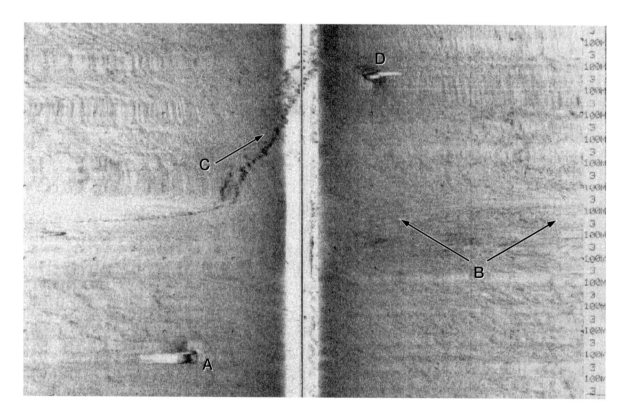

Record 60: In target search operations, timely record interpretation helps the search team to rapidly position the target. In the record above a target (A) was acquired approximately 40 meters to port by the sonar operator and the vessel was turned 180° in the direction of the target. The turn is seen in the sonar record as an apparent change in seabed configuration (B). The survey vessel's wake from the first pass is seen (C) as the vessel re-approached the target. The target was acquired a second time closer to the survey vessel (D).

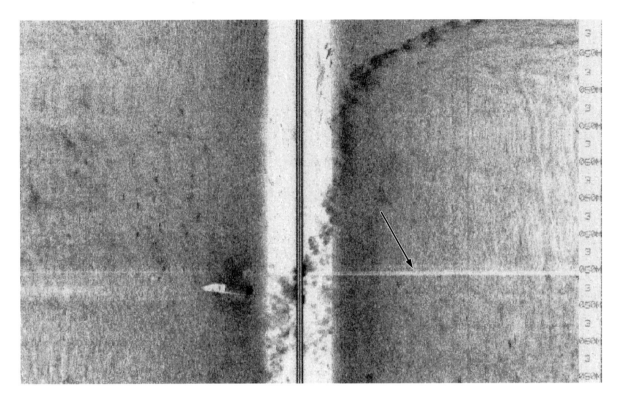

Record 61: A third pass on the target shown in the previous record is made bringing the towpath closer still to the target. Here the range is dropped to 50 meters per side. The towfish is pulled through the wake from a previous pass and signal quenching is experienced and seen as a white streak on the record. (arrow)

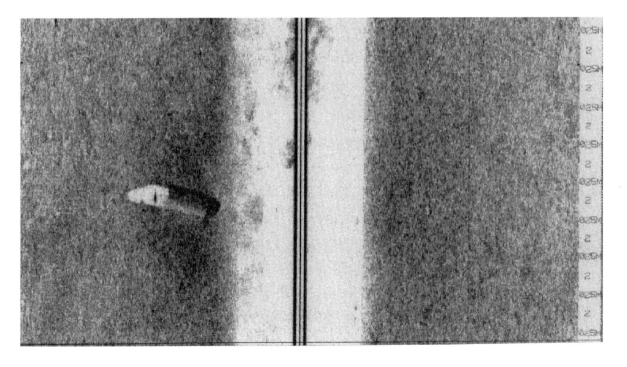

Record 62: With the target positioned accurately, the operator could safely drop the sonar range further for another pass. In this record shadow lengths and target shape can be measured accurately.

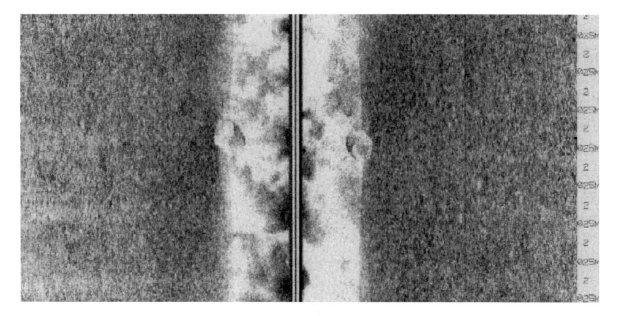

Record 63: To precisely position the target in the previous records, the sonar is towed directly over the target. As a result it is seen on both channels of the sonar. Mensuration of the target shows that it has the approximate dimensions of a 10,000 gallon fuel tank.

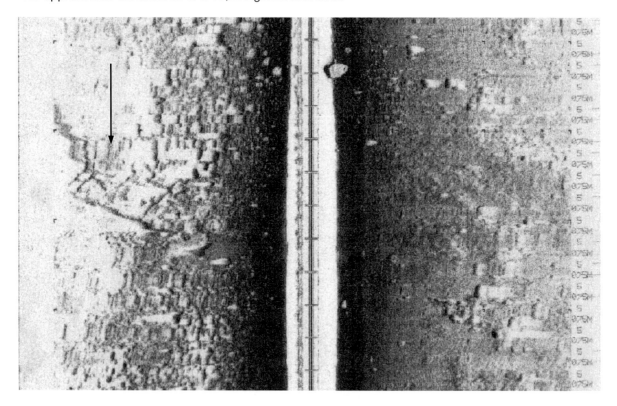

Record 64: Target search operations in rugged bottom terrain require careful attention of a skilled operator. Some of the first clues to man made targets on the seabed are the presence of straight lines on the sonar record. In the above record, some linearity is evident in the middle of the port channel (arrow). On rock strewn bottom, it can be difficult to recognize man made objects unless they posses considerable structure. Although most of the target is out of range in this record, the next pass was to port. Record 65 shows the target more clearly defined.

is shown in Record 65. Once the entire vessel is within range of the sonar, it is clearly recognized. The ladder-like objects seen on the starboard channel are typical of the return side scan receives from hull plates and ribs of steel wrecks.

One of the most important aspects of target search operations is the portion of the operational planning that deals with the selection of the search area. Even the best sonar team in the world cannot locate the target unless the area where it lies is searched. Many famous sonar operations, including a search for the Titanic, have failed because the search was made in the wrong area. The research that will lay out the proper search area saves an operational team many at-sea hours.

For historical shipwreck search, the search areas are determined by historical research and an evaluation of the bottom topography in the area of interest. For modern vessel and downed aircraft search operations, the use of computerized drift analysis is very helpful if flotsam was seen and positioned on the surface. During the planning of the search execution the maximum search area should be considered and then broken down into manageable sectors that can be completed in 1-3 days. The sectors should be listed by priority which is based on the relevant information mentioned above. This allows for a more careful analysis of the areas searched and may reduce lane lengths and shorten the amount of time it would take to resurvey any suspect lanes.

Record 65: The man made target in the previous record is shown above where the survey vessel passed directly over the wreck. The target is far more recognizable here since much more of the wreck is within the sonar range.

Chapter 8:
Record Interpretation

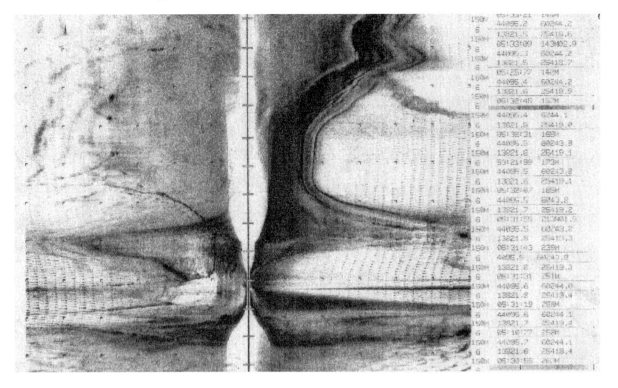

In the recent past and during early sonar development, accurate interpretation of features recorded on a sonograph were a matter of discussion and argument. This is becoming less true with the wide experience achieved by sonar specialists and the examples of the phenomena they have encountered in the form of permanent sonar records.

Even though the experience of operators is increasing, there have been efforts to make side scan sonar interpretation a function of quantitative analysis. In spite of these efforts, interpretation remains a thoroughly qualitative process. The operator must use the entire record and, often, even data recorded on earlier passes over an area, in order to accurately assess the condition of the area being scanned.

Although side scan sonar systems provide accurate images of the seafloor, there are numerous factors which add to, or detract from, the images that are generated. These factors acoustically add image components that are not represented by real components of the seafloor. Sometimes the same factors "hide" or mask parts of an image that would normally be displayed. Things which contribute to these false images of the seafloor are generated by natural objects such as fish, surface conditions or density changes in the water, as well as man-made objects such as moving ships, pipelines, wharfs and shipwrecks. Sonar images also contain acoustic shadows behind objects illuminated by the sonar beam which can further complicate the interpretive process. Good record interpretation starts with a clear understanding of the different effects caused by these factors.

Material properties of the area being scanned determine the acoustic reflectivity of that section of seabed. Rock and gravel are better reflectors than sand or mud and will therefore show up darker on the final sonar record. Further, the physical shape of the individual components of these materials strongly influence reflectivity and backscattering potential. Seabed topography also determines the reflected energy strength from the sonar beam. Up slopes facing the towfish are far better reflectors than down slopes because of the lower angle of incidence of the sonar pulse as it encounters the seafloor (see Illustration 59). Topography with a lower angle of incidence appears dark on the record. Since material reflectors and topographic reflectors often produce the same effect on the sonar display, it is up to the operator to interpret the sonar record carefully in order to determine the actual makeup and configuration of the seabed. This interpretation is made easier by understanding what kind of sonar records are generated by sand ripples, mounds and depressions, gravel, rock and other topographic features. However, since the effects of these features on the record are typically consistent and predictable, once the sonar operator has some experience, record interpretation is largely straightforward.

For timely interpretation of sonar data, it is important that the person who operates the system generating the records is the same person who is given the task of interpreting the final data. Otherwise, it would be necessary for the operator to take copious notes during record generation in order to inform the interpreter of environmental and survey related changes that might effect the data.

The following sonar phenomena is discussed with the aim of assisting the sonar operator in understanding the various effects displayed in the data caused by man-made or natural conditions. Experienced side scan operators encounter many different sonar phenomena. In every case, careful analysis of the data should result in accurate explanations for unexpected results. If the reader encounters data that seem inexplicable, very often, a sketch of the acoustic geometry and an outline of the conditions encountered during the survey will provide the clues to accurate record interpretation.

SHADOWS

The primary feature that provides three dimensional quality to the two dimensional sonar record are the shadows that occur near objects relieved from, or depressed into, the seafloor. Shadows are of extreme importance and the interpreter relies heavily on their position, shape and intensity to accurately interpret most sonar records. Shadows are one of the first clues we have of the actual conditions in a given area of seabed.

Acousticians refer to the effect of an acoustic pulse traveling across the seafloor as an insonifying or "illuminating" pulse. Any solid object that inclines from the bottom will reflect more energy back to the transducer than the surrounding area because the angle of incidence of the acoustic pulse is lower for that particular object or area. Because this object will reflect more of the total available energy back to the towfish, some of the area immediately behind the object will be insonified to a lesser degree and therefore reflect less energy. The area that reflects less energy appears lighter in the sonar display. It is this area that we refer to as the 'shadow zone'. When familiar with the common behavior of sonar, one becomes accustomed to the concept of shadow zone appearing lighter (whiter) than other insonified areas.

Shadows are not the only white areas in a sonar record. Other areas are those where little energy is being returned to the towfish by way of reflection. In an uncorrected record, the water column is an area where, assuming there are no reflectors such as fish below or around the transducers, there is little energy return. This area is often very light, if not pure white, in the record.

Many bottom targets such as a gentle upward localized slope or an acoustically translucent object cause only a slight shadow in the sonar record. Other objects such as rocks, sand waves, shipwrecks or large fish will cast a clear, harsh shadow. Thus the intensity of shadows tell us something about the makeup of the objects causing them. The cause for lighter areas on a sonar record are grouped into three general categories: 1. Shadow zones that have been blocked from the sonar beam by an acoustically opaque object, 2. Areas of topography that provide less backscattering of the sonar beam such as soft or smooth sediments, 3. Areas that are oriented in such a way as to provide less backscatter, such as an area inclined away from the towfish.

A localized depression for instance, will receive and reflect less energy from the towfish simply because the angle of incidence of the acoustic pulse is greater where the sediment slopes down. At the back of the depression, or further from the towfish, where the bottom rises back to a flat seabed, more energy will be reflected back to the towfish and that area will appear darker on the resulting record. Due to this "incident angle" related backscattering it is generally assumed that if a shadow is behind an area or object of high reflectance, that particular area projects from it's surroundings. Conversely a shadow in front of an area of higher reflectance indicates a depression in the seafloor (see Illustration 68).

A pipeline runs across the port channel of Record 69. The pipeline is small and it's acoustic reflection is difficult to recognize; the shadow it casts defines it more clearly. Where the pipeline hangs off the bottom, the shadow is cast further away and where the pipeline is buried into the bottom and has little relief, the shadow disappears. The shadow of small targets such as pipelines or cables not only help the sonar operator to identify the target but also to determine its relief on the seabed.

Since the speed of sound is reasonably consistent and ray paths are considered straight for the purposes of side scan sonar in thermally homogeneous waters, the shape of the shadows on a sonograph is usually directly related to the shape of the objects casting them. Detailed inspection of the shape of a shadow is helpful in determining the physical condition of the objects.

On the starboard channel of Record 70 there is a seven meter circular depression in the seafloor that is approximately 0.3 meters in depth. It's origin was unknown at the time of the survey, but since the shadow zone is closer to the towfish than the area of harder reflection, we know that the anomaly is a depression rather than a mound. This is confirmed by towing the sonar directly over the anomaly and examining changes in the first bottom return.

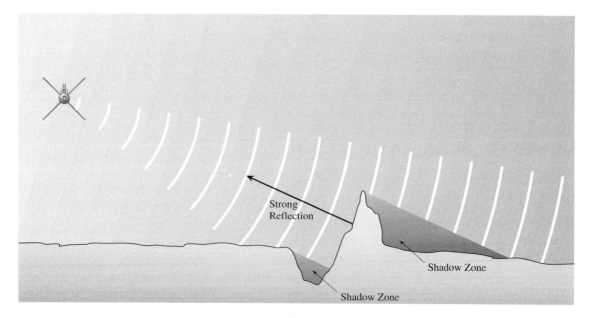

Illustration 68: On a homogeneous bottom type, shadow zones or lighter areas on the sonar record are typically a function of the amount of insonification an area receives. A shadow zone in front (towards the towfish) of a strong reflector indicates a depression in the seafloor while a shadow zone

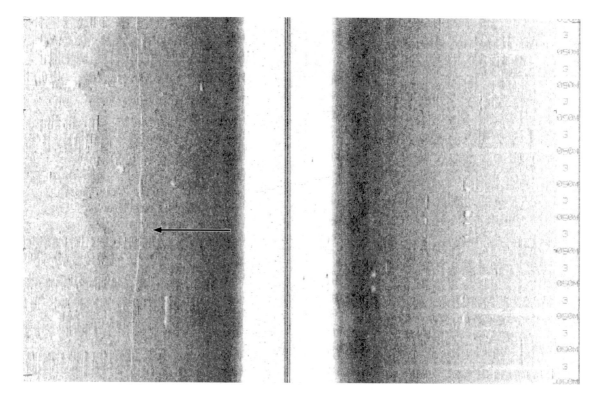

Record 69: Shadows are an important clue to a target's existence and very small targets are sometimes recognized only by their shadow. In the record above, a thin pipe is laying on the bottom and is most evident by the shadow it casts. Where the pipe is suspended above the bottom (arrow), the shadow is separated from the pipe's image.

A sloping seabed will also provide less return on one channel of the sonar than the other. In Record 71 the survey vessel passed over a bar or tongue of sediment that was shallower than the surrounding area. The outer edge or terminus of the bar was fairly steep but within the range of the sonar. This down sloping area on the port channel of the record reflected far less energy than the flatter seabed because the angle of incidence of the sonar beam was greater. The shadow zone confirms that this portion of seabed is a steep slope down and away from the surface.

The shadows cast by objects are a function of the angle at which the sonar beams strike the objects. For example, an object insonified from one angle may cast a distinct shadow, while it will cast no shadow when insonified from another angle. This can be misleading and since shadows can provide significant information about a target, objects to be identified and classified should be imaged from a number of different angles. This relationship between the geometry of towfish-target-shadow is demonstrated in Records 72 and 73.

The target is the same in both records. In Record 72 the target, which stands 3 meters off the seabed, appears relatively sharp but without much shadow (arrow). In Record 73 the shadow is far more distinct because the sonar beam is insonifying the target from a much greater angle of incidence. The lack of sharpness to the target at this range is primarily due to the effects of beam spreading (see Chapter 5 *Theory of Operation*).

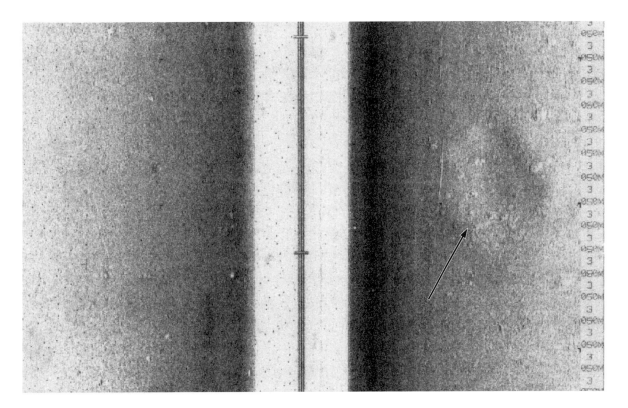

Record 70: The slight change in the seabed on the starboard channel of this record is a very shallow depression of unknown origin (arrow). The clue to the configuration of this anomaly lies in the fact that there is a weak shadow in front of the area of stronger reflection.

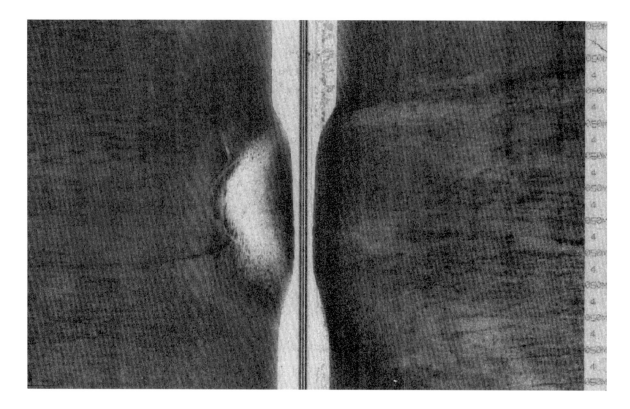

Record 71: In the record above, the sonar was towed over a sandbar-like protrusion of sediment. The lighter area on the record shows the sharp decline of the end of the sandbar away from the towfish.

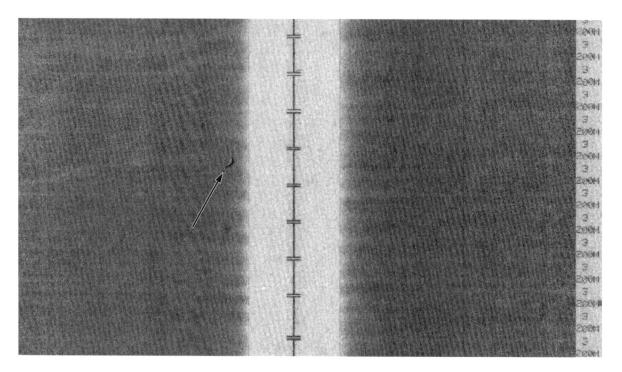

Record 72: Shadow length is a function not only the height of a seafloor protrusion but also the angle between towfish and target (sonar geometry). Objects insonified at a small angle, such as this two meter high target in the near ranges (arrow) create little shadowing. (See record 73)

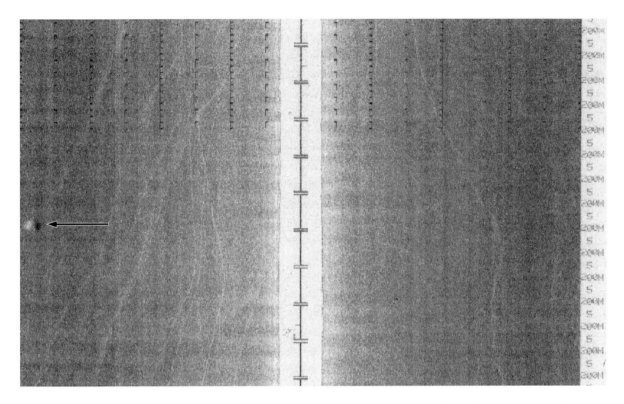

Record 73: This target (arrow) is the same as that in record 72, but imaged at the longer range of almost two hundred meters. Since the sonar beam intersects the target at a much greater angle, the shadowing is enhanced significantly. The lack of target image resolution is primarily due to beam spreading.

In uncorrected Record 74, waves of soft sand line the seafloor. These sand waves have a very steep leading edge. In this record the tow path is not quite 90° to the axis of the features and they provide good reflectors and shadow zones for the sonar. On the port channel the sonar beam strikes the sand wave's trailing face (A) at a lesser angle of incidence than on the starboard channel. This is evident by the darker image of the trailing face on the port channel. On the starboard side the sonar beam strikes the leading edge of the sand wave directly (B) resulting in a hard sonar return. On the port channel shadows caused by the sonar geometry are cast across the sand wave's leading edge.

Shadows of hard targets will often indicate to the operator considerable details about their condition. Record 75 shows the remains of a 19th century sailing bark. The keelson and it's associated shadow is seen running down the middle of the wreck. Also evident are the individual frames, or ribs, of the ship on the right hand side of the wreck. It is shown by the longer shadows, that the frames are taller at the lower right side of the wreck. There are similar frames on the opposite side of the wreck but the angle between them and the towfish is high enough to prevent pronounced shadows. Lesser insonified areas such as these shadows provide further indication of the existence of strong reflectors and will even allow us to see individual objects, such as the ship frames, more clearly.

The corrected Record 76 is of the wreck of the *Larchmont* sunk off the coast of New England in 1907. In the sonar record, at least two boilers are visible and one of them still has the remains of a stack standing on it. The remains of a paddle

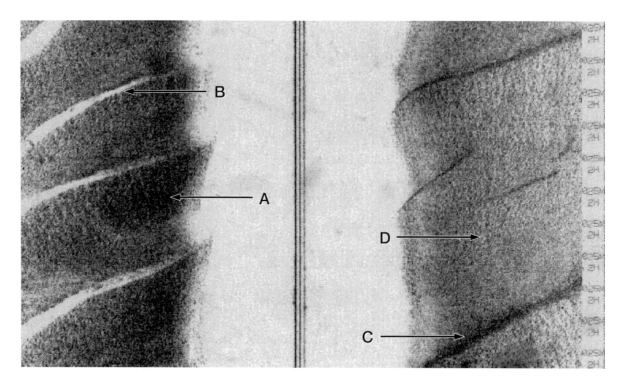

Record 74: Steep sediment waves will cast more or less shadows depending upon the angle at which they are insonified. This record demonstrates a number of different effects of sonar-seafloor geometry. The sand waves in the record above are similar on the port and starboard channel. The gentle up slope of the wave backs (A) on the port channel form an inclining surface to the sonar beam (good reflectors) while the wave fronts (B) create shadow. The wave fronts (C) on the starboard channel are inclining in relation to the towfish and are reflectors while the wave backs on this channel (D) slope away thus showing up lighter in the record.

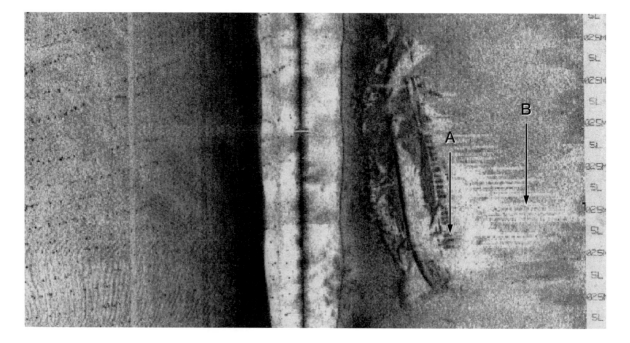

Record 75: Shadow formation can show considerable detail of the objects casting them. Here the frame tops (A) of a sunken Bark cast individual shadows (B) that relate directly to the frame heights and spacing.

wheel box and the structure within is evident and in the shadow of the paddle wheel, the filigree pattern of the paddle box cover is clearly visible. Here, as in many cases, the shadow in a sonar record provides the operator with a far better concept of the condition and substance of the target than the direct reflections. By comparing the sonar record with a photograph of the vessel in it's original condition (Illustration 77), the interpreter can determine the bow from the stern of the wreckage by seeing the position of the stack in relation to the paddle wheels.

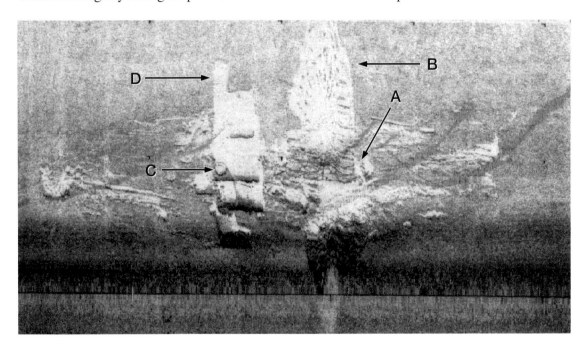

Record 76: Shadows play an important part in identifying the orientation of the shipwrecked paddle wheel steamer *Larchmont* on the seabed. The wreck is heavily deteriorated and it is difficult to determine bow from stern. A is a paddlewheel. B is the shadow of the paddlewheel covering. C is a stack still standing on one of the boilers and D is the shadow of the stack. See Illustration 77.

Illustration 77: In comparing this photograph of the *Larchmont*, with Record 76, we can see that the paddlewheels were positioned aft of the boiler stacks. This provides clues to the wreck's present orientation on the seafloor. *Courtesy William P. Quinn Collection*

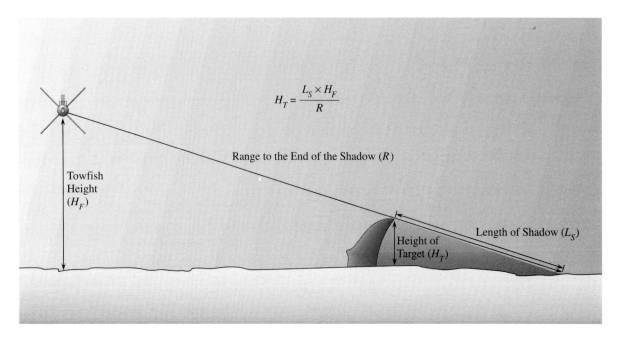

$$H_T = \frac{L_S \times H_F}{R}$$

Range to the End of the Shadow (R)

Towfish
Height
(H_F)

Length of Shadow (L_S)

Height of
Target (H_T)

Illustration 78: The height of an object in contact with the seafloor can be determined by measuring ranges and shadow length and applying the formula as shown above.

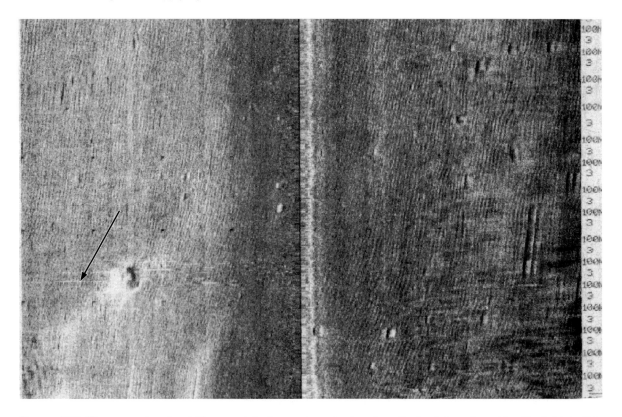

Record 79: The operator should be careful when using shadow lengths to determine object height. In this corrected record, animals swimming in the water column have contributed to the apparent shadow length of a target on the port channel (arrow).

Since the acoustic path from the sonar is relatively straight, there is usually a right triangle formed with the three angles at the towfish, the seabed and the tip of the target shadow. The target lies along the base of this triangle with its highest point intersecting the hypotenuse. This geometry forms two similar triangles and, by using a mathematical ratio, the height of the target is calculated.

This geometry is shown in Illustration 78. The calculation used to determine the height of a target is as follows: the height (Ht) of an object is equal to the product of the acoustic shadow length (Ls) and the height of the fish (Hf) above the seabed, divided by the range to the end of the shadow thus:

Target Height = (Shadow length x Height of the Fish) / Range to the end of the shadow

This calculation is very accurate in normal sonar operations, however care must be taken when operating in unusual conditions. If there are severe density differences in the water column, acoustic ray paths are not necessarily straight and if severe, the height calculation from shadow length may not be accurate. Further, if there are other targets in the area that affect the primary target's shadow, care must be taken to be certain that the proper geometry is being applied. In corrected Record 79 the shadow of a school of fish in the water column are overlaid on the shadow of a target at longer range (arrow). This could cause confusion about the primary target's shadow length unless the operator is careful to distinguish between the two shadows.

When imaging complex targets, the exact position of the target component casting the longest shadow may be difficult to determine. In cases where target height must be calculated conservatively, the interpreter should assume the shadow causing component to be at the leading edge of the target (or that edge of the target closest to the record centerline).

Sonar geometry can prevent a target from casting a visible shadow on the seabed. For instance, a target that is in the water column at a height similar to that of the towfish may cast a shadow that is beyond the range being displayed. Further, if an object is above the towfish, while the targets image will be displayed, it will not cast a shadow on the seabed.

TOPOGRAPGHY AND GEOLOGY

Side scan sonar is an ideal tool for determining the configuration of sea, river or lake beds over a large scale (100's of meters). For larger scale (>2 kilometers) area imaging, although resolution is sacrificed, lower frequency sonar systems are used. In order to maintain resolution at large scales, mosaics of 100 kHz sonar records are made. However, often, many single pass seabed records provide the data required in determining topographical configurations.

Sediment "waves" occur in many places underwater particularly where there are current velocities sufficient to move sediment particles. In Record 80 both large (mega) and small (micro) sand waves are evident on the seabed. For regular sediment waves the wavelength is calculated by determining the crest to crest distance. In the record above, the mega-ripples are about 130 meters in wavelength.

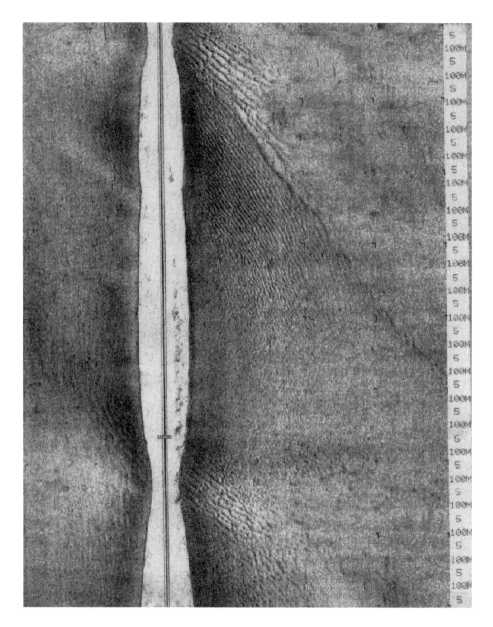

Record 80: Topography can be assessed more thoroughly using side scan sonar than with bathymetrics alone. In the record above, small sand waves lay on top of much larger ones which have a wavelength of 130 meters.

In areas where the open seabed is obstructed by a shipwreck or other obstacles, sediment transport is disrupted. Around these obstructions, sand waves of varying sizes, heights and direction indicate large fluctuations in current velocity and direction. This can be seen in Record 81 around a steel shipwreck sunk over seventy years ago. The wreck lies on a shoal over which tidal current velocities reach 3.0 knots. In this case the predominant sediment flow is to the west, thus resulting in a significant buildup of sand on the east side of the vessel. Wrecks can disrupt normal sediment transport over hundreds of meters (see Record 136 *Monitor*).

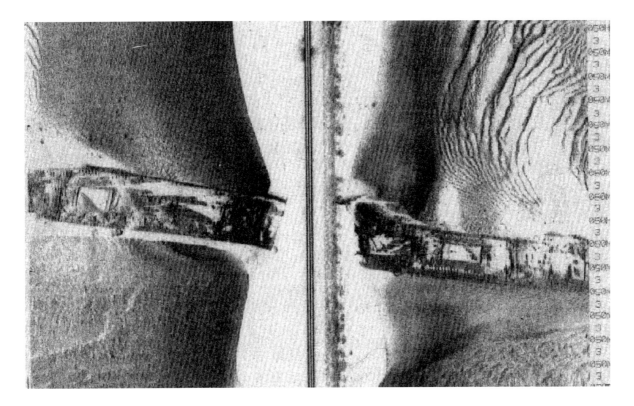

Record 81: Local fluctuations in current velocity can significantly disturb sediment transport in a region of seabed. Here, in the vicinity of a large shipwreck, the alterations in current direction and velocity have formed features known as 'wreck marks" in the bottom topography.

Record 82: A one meter high ledge off the starboard side of the towpath is a hard sonar reflector. It is clearly displayed as a dark line (A) on the starboard channel. The hard reflection from the ledge face is caused by the change in angle of the seabed. The sonar beam's larger angle of incidence at the top of the ledge where the seabed levels out again causes a lighter area on the record (B).

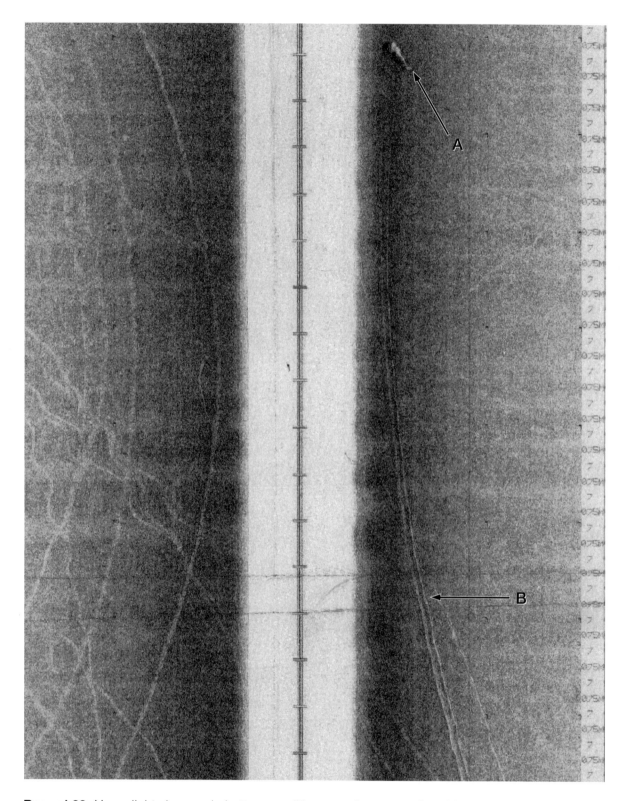

Record 83: Very slight changes in bottom conditions can be seen using side scan. The object on the seabed (A) has been moved along the bottom by a fishing dragger. The sonar team could follow the drag marks (B) back to the point of origin by towing the sonar on a track parallel to the marks.

Small shoals or ledges are easily delineated by side scan. In Record 82 a small ledge, approximately one meter in height, is seen on the starboard channel. The light shadow zone beyond the ledge is due to the change in incidence of the sonar beam on the seabed at the top of the ledge.

Even small marks on the seabed are well defined. In Record 83 the long parallel lines curving across the port channel are from trawl doors towed along the seabed by fishing draggers. The small target at the top of the record is the fuselage of an airplane which has been moved along the seabed. The surveyor could follow the drag marks of the plane to locate the area of its origin.

In the corrected Record 84 there is evidence of fishing activity in multiple directions on a flat seabed. Most dragger marks occur in pairs because two doors are used in this common method of fishery. Low, normal and high contrast settings were used when generating this record. These contrast levels are noted in the data channel next to the gain numeral. When normal contrast is used, the gain numeral stands alone.

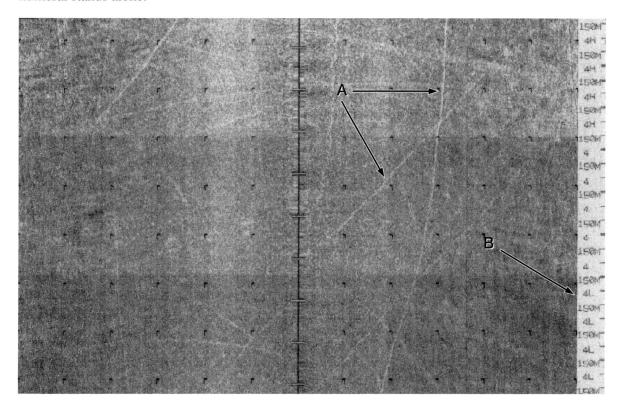

Record 84: As small disturbances in the bottom topography become smoothed by currents other natural processes they become faint but remain detectable by sonar for some time. The lines (A) on this record were created by fishing draggers several months prior to the bottom survey. The recorder printed this record set at low, normal and high contrast as indicated by 'H' or 'L' in the system status channel on the right (B).

TURNS

Vessel turns while towing a side scan sonar system have effects that are readily seen in the sonar data. Turns will distort what we think of as a normal record. During a turn, the sonar transducers no longer scan the seabed in a consistent, straight line. The transducer on the inside of the turn is insonifying and recording reflections from a much smaller area of the seabed than the transducer on the outside. When these images are observed on a straight display, they can be seriously misleading. During target search, turns provide images that resemble targets. During general area survey, turns provide images of geological features that do not exist. The operator should not rely upon interpretation of records generated during turns and it is important to recognize the effects that turns have on sonar data.

In Record 85 the survey vessel is towing the sonar across a seabed covered with sand waves. A 90° turn to port brings the tow path parallel with the wave crests. Although it appears that the geology has changed, in reality it was the vessel track that changed.

The towfish altitude also has a tendency to decrease when making a turn. While towing, the drag of the towcable in the water column imparts a significant upward force on the towfish. In normal operations, water mass movement relative to the towcable is balanced against the amount of cable out and the weight of the towfish which maintains fish height above the bottom. Turning reduces the water mass movement by the cable and has the same effect as slowing the towing vessel resulting in the towfish dropping. This effect increases dramatically with decreasing turning radii and with increased lengths of cable in the water. Increasing the speed of the survey vessel during turns is one method of counteracting reduction in drag. Turns executed when using a depressor will cause a less pronounced loss in towfish altitude.

Small navigational corrections are also considered turns and effect the sonar data. This is particularly noticeable when scanning a small target. If the survey vessel makes small course changes while countering wind or tides, in order to traverse a reasonably straight line, these course changes may translate to the towfish.

In the uncorrected Record 86, a small ship wreck is being scanned. The survey vessel passes directly over a portion of the wreck and the detail is good. The shadow of the target is also clear and provides further information. However, the wreck is not curved as it appears in the record. Although the seas were calm, the survey vessel captain made minor course changes while scanning the target. This caused the towfish to change it's path through the water and induce curves in the image that do not exist in the actual target.

During target search operations, false image generation during turns can be particularly baffling. Record 87 was made on a turn between survey lanes during a search for a downed aircraft. The turn to port over a rocky bottom, resulted in the sonar printing an image remarkably consistent with that of an aircraft (arrow). Only a re-scan of this portion of seabed convinced the search team that an airplane did not exist in this location.

Although the tendency for many vessel operators is to make course changes during target imaging, it is very important not to make any maneuvers that will turn the towfish while the target is being scanned. Turns made during large-scale survey operations may not appear to affect the data significantly, however, they can cause apparent changes in topographic structure.

Record 85: The sonar operator needs to be aware of any course changes made by the towing vessel since turns can severely effect the data. In the record above, the survey vessel turned 90° to port generating an apparent change in the bottom topography.

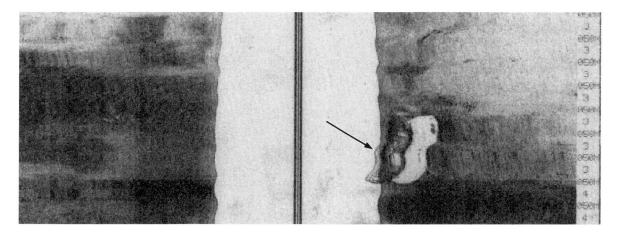

Record 86: Targets, as well as bottom topography, will reflect changes in vessel course as can be seen in this record of a sunken vessel (arrow). The curves in the target are caused by very small course changes in the survey vessel's track followed by the towfish.

Record 87: During a turn, the transducer on the inside of the turn, images targets for a longer period of time than the opposite transducer. This phenomena can create unusual images such as that of the airplane shaped shadow (arrow) on the port channel of the above record.

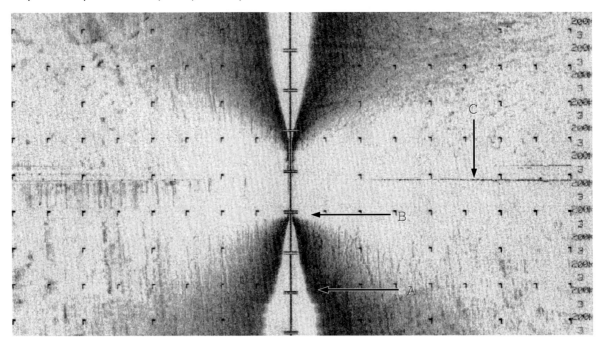

Record 88: If the towfish is lowered too deep, it can strike the bottom and be damaged or lost. In uncorrected records, the first bottom return approaching the trigger pulse (A) is a warning of potential collision. When running corrected records, an altitude alarm will sound, alerting the operator to check the towfish height. On the record above, as the towfish approaches the seabed the record becomes lighter (B). This is a result of the increasing angle of incidence of the sonar beam. The dark streak (C) is noise created by the friction of the fish striking a sand bottom.

TOWFISH-SEABED COLLISIONS

One of the hazards when towing the subsea transducer array on long cables is the danger of hitting the bottom. This can be damaging to the towfish and should be avoided. In Record 88, too much towcable was deployed allowing the fish to fall down onto smooth sand. Note the loss of signal return as the towfish dropped towards the bottom, bringing the angle of incidence of the sonar pulse higher and higher until finally, with the fish in contact with the bottom, there is almost no bottom reflection. Even in a smooth sandy sediment, the shock of the towfish hitting the bottom causes spikes of noise on the record.

In Record 89, the fish collided with the bottom because the survey vessel towed it over a rock ledge. In this record the noise spikes are more noticeable because of the contact of the fish with rock.

Record 89: When the towfish strikes a rock bottom, damage is more likely than when it collides with sediment. In the record above, the towfish hit a rock ledge (A). In this case the survey vessel towed the sonar into dangerously shallow water. When the towfish struck the rock ledge, friction resulted in multiple noise spikes (B) on the record.

NOISE

"Noise" in terms of underwater measurement, is defined as a signal which we don't want. At times, particularly with older side scan sonars, noise will occur in the resulting data that is generated within the system. The noise might appear as flecks of dark speckles, streaks, continuous lines or bands such as those seen in optical video aliasing. When these noise problems occur, changing subcomponents or power sources, rerouting cables or resecuring the system ground often helps.

Most of the noise that modern sonar systems sense and display, however, comes from the underwater environment. Much of this externally generated noise is recognizable as such, thus relieving the operator from unnecessary trouble shooting.

We often think of the underwater environment as a relatively quiet place. Divers know that when water covers their ears, very little sound is heard. However, the sea is full of a wide spectrum of ambient noise out of the range of human hearing. Some noise is generated by the effect of wind on the ocean's surface. Rain, ships propellers and currents also produce noise. In tropical waters, animals that bite and crush small pieces of coral while feeding on the polyps create noise. Although one or two fish would not contribute significantly to the ambient noise level, thousands of fish in a relatively small area create a virtual din. Some species of shrimp also create noise as they move through the water. These "acoustic generators" make the ocean a difficult place to work when they interfere with our own generated sound signals.

When the outgoing pulse from a sonar system strikes a target, it doesn't merely come back to the towfish and disappear. Portions of the output pulse reflect back and forth between the seabed and the surface of the ocean many times. Usually the intensity of these echoes is small enough that they are not displayed on the sonar system. However there are many other noise generators in the ocean that we do sense and their effects are readily seen in a side scan sonar image. Since noise is an undesired signal, it is important to recognize it when it occurs.

Most side scan sonar systems are sensitive to the high acoustic output of ship propellers but since the transducers on the sonar are very directional, propellers from the towing vessel rarely affect the sonar operation. In Record 91, propeller noise from another craft, traveling in the opposite direction, is seen as a dark streak at the outer edges of the record. The band of noise data is narrow because the transducers are very directional and they, therefore, receive signals from a very narrow section of the ocean. Propeller noise, as many types of acoustic interference, is most often evident at the outer fringes of a sonar record because, when this portion of the record is being generated, the time varied gain is set very high.

In Record 92 another craft's propeller is seen on the port channel. The other vessel is out of sonar range and traveling in the same direction as the survey vessel. The persistence of this noise, displayed over time, indicates that the other vessel is maintaining position relative to, or traveling in the same direction as, the towfish.

In Record 93, the survey vessel passed another craft to starboard traveling in the opposite direction approximately 65 meters away. With the sonar range set to 100 meters, the other vessel's track is within the displayed range of the sonar. It's wake is seen in the sonar image after the vessel has passed. The range to the other craft cannot be determined by the noise spike (arrow) because it is generated independently from the sonar system. In this record, the range is only indicated by the position of the vessel's wake. Another wake, made by a vessel which passed earlier, is seen on the opposite sonar channel.

Direct propeller noise is often easy to identify. Less obvious is low level drive train noise that may be recorded only when the sonar is running at long ranges. Rhythmic drive train noise is seen in Record 94. Although similar in appearance

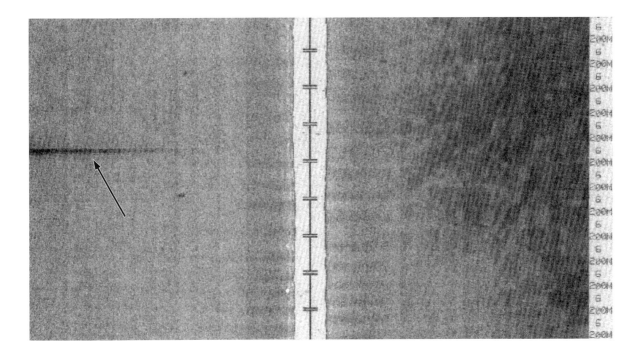

Record 91: The sonar towfish receives it's acoustic signals from a very narrow zone as it is towed through the water. The propeller noise (arrow) from a vessel passing in the opposite direction appears only briefly on the record because the other vessel is within the sonar receiving zone for a short time. Since time varied gain circuitry is set high at the end of each long range scan, externally generated noise is most evident at the outer edges of a sonar record.

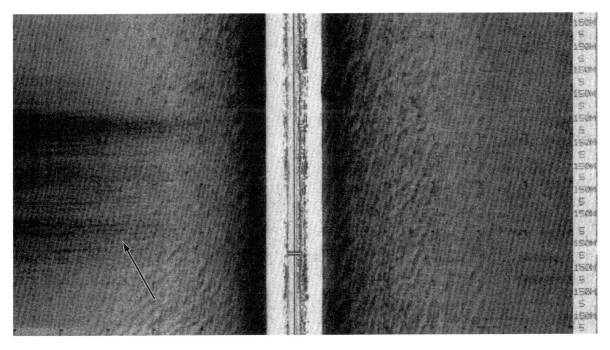

Record 92: If another vessel overtakes the towing vessel traveling in the same direction, it's propeller, as a prime noise generator, may be in the sonar's receiving zone for a long period of time as seen in the above record (arrow).This noise display, overlaid on the seabed image, can mask bottom targets.

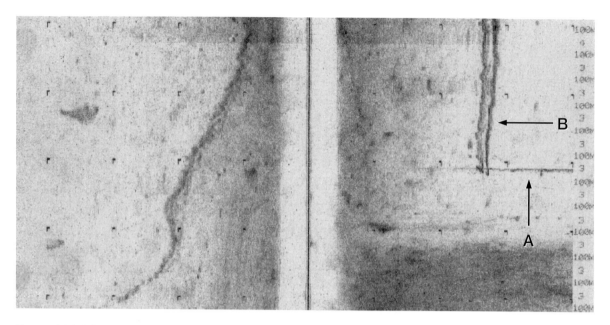

Record 93: The above record was generated in a busy coastal waterway. A power boat, traveling in the opposite direction, passed the sonar towfish at a range of approximately 65 meters on the starboard side. The associated propeller noise (A) and wake (B) can be seen on the starboard channel. Note that, although the range to the other crafts propeller is 65 meters the noise spike still appears at the outer range of the record.

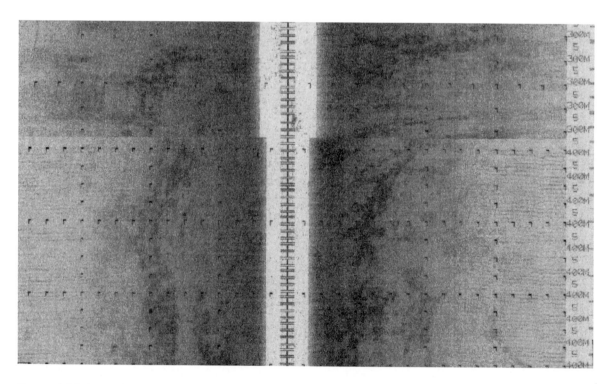

Record 94: Less obvious than direct propeller noise, survey vessel drive train noise may be sensed by the sonar system when operating with short (< 100 meters) cable lengths. Increasing the horizontal or vertical distance between the survey vessel and towfish will often eliminate the noise. In this record, the sonar range was reduced and the noise elliminated.

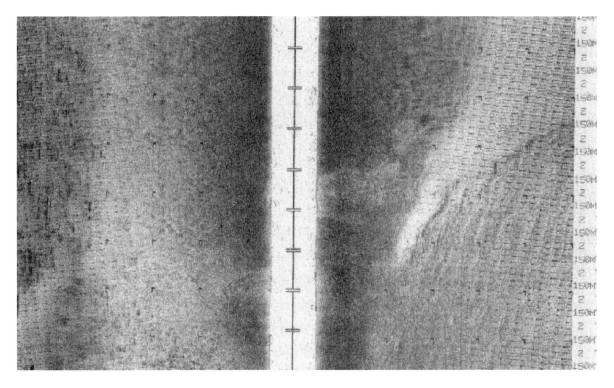

Record 95: Bottom profilers, sparkers and other timed sound sources can interfere with side scan sonar. These interference patterns are typically seen as regularly spaced dark lines on the sonar display. As in the case of most externally generated acoustic interference, it is most evident at the other fringes of the sonar record. The interference in this record is caused by a bottom profiler and is more evident on the starboard channel because the interfering instrument was positioned to the starboard side of the side scan towfish.

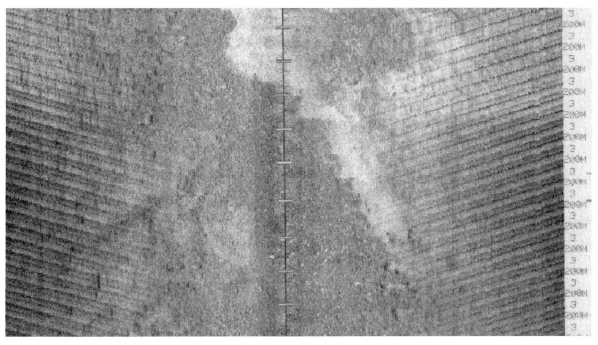

Record 96: Interference patterns from other instruments can be mild as in Record 95 or severe enough to mask potential bottom targets or structure as in the above record. Increasing the range between the sonar and other instruments will reduce the effect of the interference.

to sub bottom profiler noise, the faint lines on the 400 meter ranges here are from the survey vessels own propulsion system. Although the noise is not particularly objectionable and probably would not mask large target signatures, shortening the ranges from 400 meters to 300 meters removed most of the noise from the record. Other methods of reducing the interference of the survey vessel's propulsion system include lowering the towfish to increase its vertical distance from the boat, or winching out more cable to increase its horizontal distance from the boat.

Sparkers, boomers, sub bottom profilers, as well as bathymetric profilers and depth sounders can all generate 'noise' on the sonar display. A continuous and regular interference pattern of noise may overlay the sonar display when depth sounders, seismic instruments or other timed sound sources are used locally, and within the frequency band of the side scan sonar. The pattern that these external instruments make on the side scan display are a function of their pulse rate and the display update rate (range setting) of the side scan sonar. It is this characteristic that makes interpretation of these signals reasonably direct. In Records 95 and 96 the same profiler is in operation leaving a noise pattern on the sonar record. The noise is most noticeable on the starboard channel of the sonar record because the other instrument was positioned to that side of the side scan. The difference in the noise pattern between the two records is a result of different range settings of the side scan. Not unlike other faint noise sources, these signals are displayed most strongly at the outer ranges where the TVG gains are set high. These signals do

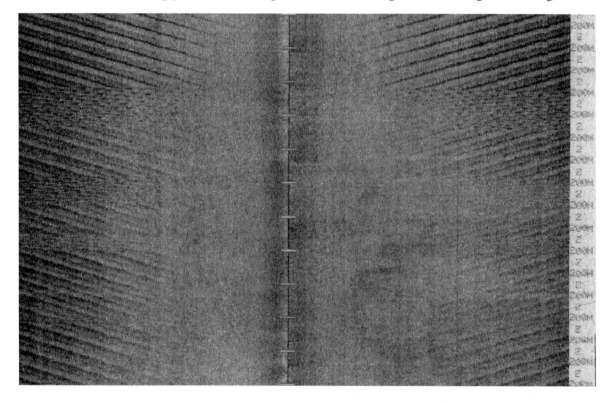

Record 97: Chart speed changes as made in the above record will alter the printed pattern of interference from timed sound sources operating independently of the side scan sonar system. This is one way to determine if interference is internal or external to the side scan recorder/printer.

not cast shadows and are usually not dense enough to mask seabed targets. Also, as with drive train noise, repositioning the sonar towfish or the transducers of the interfering instrument will often minimize this interference.

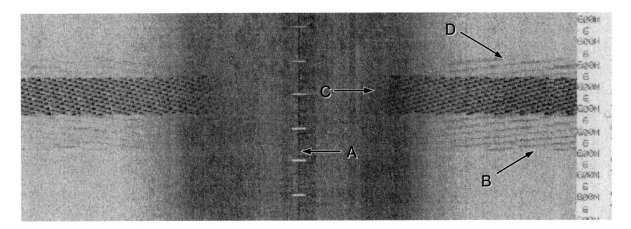

Record 98: Because of the TVG circuitry, when running extremely long ranges sonar systems are more susceptible to external noise interference. In the record above the survey vessel increased speed at (A). The increase in speed brought the towfish higher in the water column where it began to receive rhythmic drive train noise from the survey vessel (B). As a check to determine the source, the ship's profiler was turned on (C). This created a very strong interference pattern indicating that the original noise did not come from that instrument. With the profiler turned off, the vessel speed was reduced bringing the fish deeper and out of range of the noise source (D).

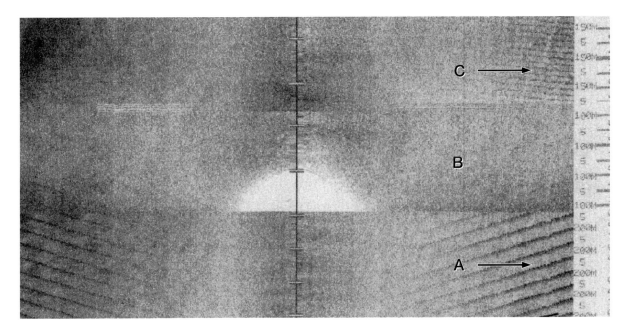

Record 99: Another method to minimize the effect of acoustic interference is to change scale setting of the sonar system. With the same towfish-profiler geometry, varying effects can be seen in the above record. Set at 200 meters the sonar system displays acoustic interference (A) . When the sonar setting is reduced to 100 Meters the interference is not evident (B). With the system set at 150 meters the external acoustic interference is only slight (C).

Chart speed changes on the side scan recorder will affect the pattern of profiler interference. The sonar operator may use this fact as a diagnostic means to determine the source of the noise. In Record 97 the chart speed was changed as a quick check to see if the printed pattern was caused internally or externally to the sonar system.

During searches for large objects using long sonar ranges, these interference patterns printed on the sonar record sometimes prevent bottom targets from being recognized by the operator. Noise should be identified and eliminated by the operator if it is felt that it may mask bottom configurations on surveys or prevent target recognition during search. If, for example, the target of a shipwreck search is a large vessel but sonar contact is made on the face of the bow or stern, noise may mask the target image.

Other methods of troubleshooting noise on the record is to change vessel speeds. In Record 98, the ship speed was increased causing drive train noise. The operator asked the bridge if the profiler was running. As proof that it had not been running, the profiler was turned on. The operator then realized that the original noise was caused by the increase in vessel speed which brought the towfish higher in the water column and vertically closer to the survey vessel. The profiler was turned off and when the vessel was slowed, towfish depth increased and the drive train noise disappeared.

The choice of range settings during sonar operations may effect whether a profiler will interfere with the sonar. In Record 99 the same profiler has more or less effect on the record at sonar ranges of 100, 150 and 200 meters. For small object detection and recognition, the increased profiler interference at long range settings, as well as the smaller displayed image size may prevent the operator from recognizing the target. Shorter ranges would be a better choice in this situation.

Noise from natural sources also effect the quality of sonar data. Natural noises are often distinguishable from measuring instruments by their pattern. Marine mammals may mask the entire record with interference. This phenomena is commonly encountered in temperate and tropical oceans. Mammals will occasionally come within range of the sonar and emit sounds which are annoying to the sonar operator. In Record 100, porpoise were emitting their own sonar signals near the side scan towfish. Porpoise emit sonar signals in the 90-100 kHz range which will be sensed by side scan receiving electronics. In this case, the noise is spread evenly over the record and appears in bursts on both channels simultaneously. The port channel is not tuned to print the seabed but in the starboard channel a targetless shadow is seen. This type of shadow is indicative of an animal or other target in the water column. Because porpoise breathe air, and swim near the surface, a check of the surroundings will often determine if there are porpoise near the towfish.

As mentioned, the ocean has its own noise level from waves, currents, rain and ships at distances. We refer to these ubiquitous rumblings as *ambient noise*. In lakes and other enclosed bodies of water, ambient noise is low but in most oceans it is significant. We have seen from the records of propeller noise that when the side scan sonar receives its echoes from longer ranges, the TVG is set high. This high amplification makes the system sensitive to very small levels of ambient noise. At long ranges, bottom reverberation from the sonar's outgoing pulse is not always strong enough to be sensed by the system over other sounds that exist in the ocean and internally generated system noise. Under these conditions, the sonar system will register and print these noise levels at the outer edges of the scan and they usually appear similar to far range seabed reverberation.

If a target reflects enough energy back to the towfish it will still be detected at long ranges but it's related shadow will often be filled with the detected ambient noise. As a result, when working at long ranges with sonar, a large target projecting well off the sea bed, will frequently have no shadow. In Record 101, a target near the 300 meter range is shown (A). The target appears as a jagged line with no shadow. The target image is not straight as a result of fish instability and thermal changes within the area scanned. If this image was seen in the near ranges, it might be rejected because of the lack of a distinct shadow. However, this target, which stands several meters off the seabed, does not demonstrate a clear shadow due to the ambient noise level filling this portion of the record.

The same target is scanned closer to the towfish (Record 102). The result is that the target has a much more distinct shadow. The shadow is present on the record because, at the shorter ranges, the TVG is not set high enough to register the oceans ambient noise or self generated system noise. The target also appears to be straighter, since the thermal effects are not as pronounced with the shorter acoustic

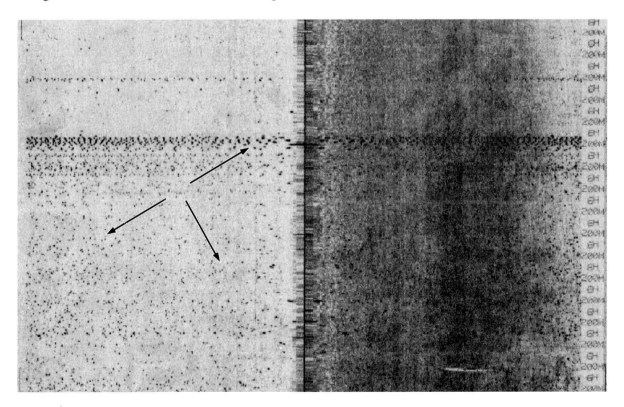

Record 100: Acoustic interference from natural sources may be sensed and displayed by side scan sonar. Since natural noise is not from a timed source, the pattern it creates on the sonar display is random in appearance rather than even and regularly spaced. In the above record, porpoise emitting acoustic pulses in the 90-100 kHz range, swam near the towfish. The resulting noise received by the towfish is evident in bands and specs on the sonar record (arrow).

path. Because of the lack of shadows behind far range targets, the surveyor should not rely on shadowing as an important interpretive tool on the outer edges of long range records.

Record 101: The sonar display may not show shadows associated with detected targets because the system receives and displays ambient underwater noise. In the above record a 4 meter high target (arrow) imaged at 300 meters on the port side has no shadow. The shadow area is filled in by ambient noise (see record 102).

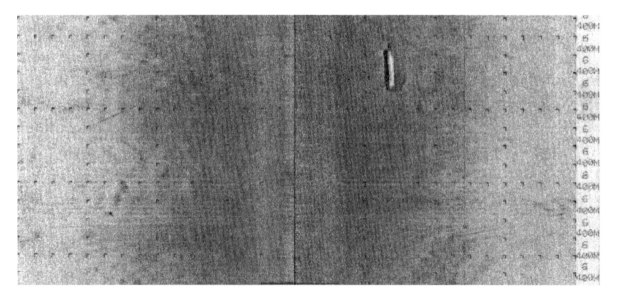

Record 102: In this record the same target as in record 101 is shown at a range at 125 meters. At this range, the target displays a distinct shadow since the TVG is not yet high enough to amplify ambient noise.

WATER DENSITY

Side scan sonar works well in many applications because of the straight path that sound takes underwater. This is true in homogeneous waters where most side scan operations take place. But sonar beam paths are affected by non-homogeneous water which may be encountered during sonar surveys.

If the velocity of sound were the same at all places underwater, sound pulses would travel in straight lines making the interpretation of sonar records easier. Unfortunately, it is not. The velocity of sound is different at different depths, temperatures and salinities. Particulate matter and bubbles also effect sound velocity but these are rarely of high enough concentration in the water to cause concern for side scan sonar operations.

The boundary between materials of different densities acts as an interface to most wave based energies. In water, salinity and temperature are the two most common causes of varying densities with the latter being more frequently encountered for the side scan operator.

Any transmitted sound pulse, such as one from a side scan sonar, tends to bend towards the water of the lowest temperature, salinity or pressure depending upon which factor predominates. Therefore, it tends to bend away from water of high temperature, salinity or pressure. This phenomena, called *refraction* or *acoustic ray bending*, occurs with side scan whenever marked density discontinuities in the transmission medium are encountered by an outgoing or reflected sonar pulse.

The most noticeable refraction in coastal side scan sonar records occurs during the warmer months of the year when the sun heats the upper water layer and this is where the side scan transducer assembly is often towed. However, this phenomena can occur in coastal waters at any time of the year and should be anticipated by the sonar operator even in areas where water column mixing is thought to be complete.

In the region where Records 103, 104 and 105 were made, a horizontal *thermocline* between a warm upper layer of water and colder deep water caused refraction of the sonar beam. These records are classic examples of coastal zone refraction. The outgoing pulse is refracted downwards when it encounters the thermal discontinuity at an oblique angle. The effect directs an inordinately large amount of energy, which would normally propagate out to the edges of the scanned range, down to the seafloor. This is part of the reason that bottom reverberation is very weak beyond refraction bands. The effect appearing on the sonar display is a thin dark band where the bottom is heavily insonified causing reverberation to be localized. The sawtooth effect noticeable in the bands of refraction is a function of a number of variables including the angle of incidence of the sonar pulse to the thermal layer and changes in the spatial relationships between the towfish, the thermal layer and the seafloor.

In Record 103, very mild refraction due to temperature variations in the water column on the starboard channel is seen. This level of refraction does not degrade the image significantly but is indicative of possible problems if survey lanes progress to the starboard. In Record 104 stronger refraction bands are seen. On the starboard channel at a range of 50 meters a target casts a long shadow. Mild refraction might not mask a target of this size, therefore the refraction could be tolerated. However in Record 105, the refraction is severe enough to be range limiting. The refraction here, caused by a marked thermocline visible on the port channel, will mask even very large targets. When encountering severe thermoclines during sonar operations, lowering the towfish or shortening the range are often the only solutions to this type of image degradation.

In Record 106, the effect on refraction of raising or lowering the towfish is demonstrated. In this record, survey lanes were being run at a range of 300 meters per side with the towfish positioned approximately 35 meters off the bottom. There is some mild

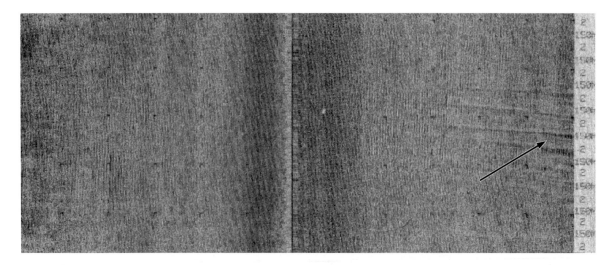

Record 103: Temperature differences in the insonified water column affect the ray path of the sonar beam. Known as refraction, this effect can be seen on the starboard channel in the above record (arrow). Jagged dark and light bands such as these are a classic indication of refraction in warm-over-cold water layers.

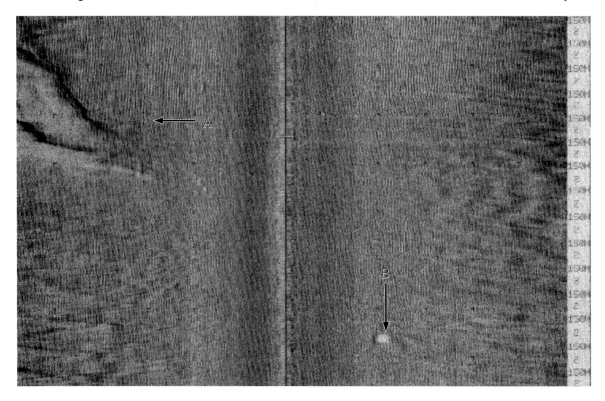

Record 104: When temperature gradients become severe, refraction degrades a sonar image significantly (A). The dark and light bands on the record can hide small targets such as those at (B) and such anomalies may go undetected within refraction bands.

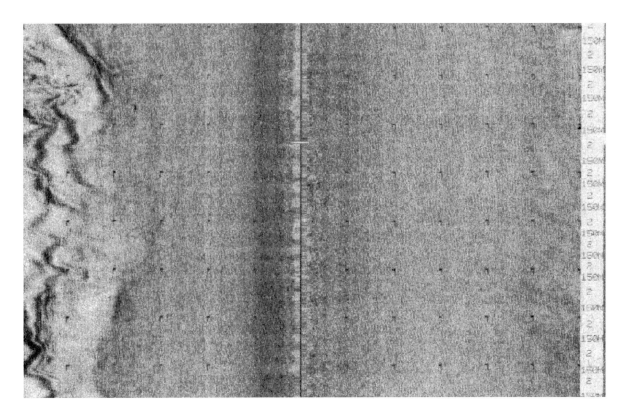

Record 105: Severe refraction such as that seen on the port channel can be significant enough to limit the useful range of the sonar.

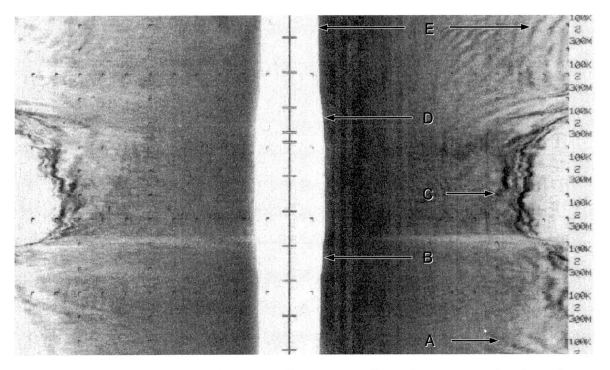

Record 106: Modifying the towfish-seafloor-thermocline geometry will sometimes overcome the effects of range limiting refraction. In this record mild refraction is visible at (A). At position (B) the towfish was raised higher in the water column. This is seen by the increased range to the first bottom return. With the towfish higher in the water range limiting refraction becomes severe (C) but is reduced when the fish altitude is decreased (D) and virtually disappears as the towfish altitude drops (E).

refraction evident at the longer ranges (A). Since the search was for a large target, this level of refraction was tolerable. From point B to point C on this lane, the towfish was raised in the water column to a height of approximately 40 meters. The refraction resulting from this altitude change is clearly seen.

When the effects of refraction become range limiting on the sonar system, and the problem is not solved by lowering the fish, running on shorter ranges may be the only solution. The operator must recognize the range limiting effects of refraction to prevent seabed areas from being uninsonified during surveys.

Acousticians will explain that ray bending due to thermal effects are either very noticeable (Record 105) where the thermocline is a steep temperature gradient, or they are very mild and subtle as in Record 107. In this record, a thermal inversion sometimes known as "winter thermocline" has affected the sonar beam. Looking very much like an underwater shoal, the boundary displayed on both channels is a result of a mild thermal discontinuity in the water column. A slightly warmer layer has been introduced under a cold layer of seawater. This condition bends the sonar beam up towards the surface as opposed to the effect in Records 105 and 106 where the beam was refracted downwards.

Unlike deep colder layers where there is a steep gradient in temperature, thermal inversions are typically not range limiting and allow enough acoustic energy to propagate beyond the thermal discontinuity to insonify the seabed at the outer ranges. Indentations from fishing dragger doors are displayed at the outer ranges of Record 107.

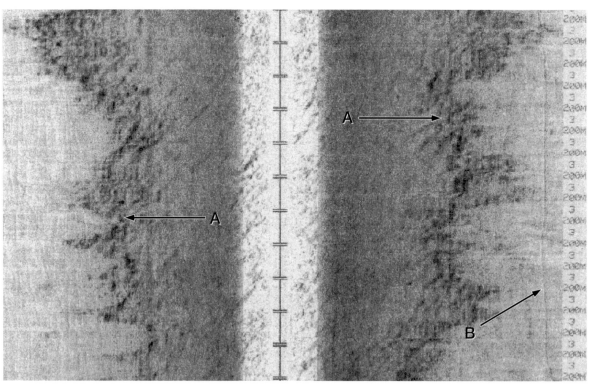

Record 107: Temperature inversions of warm-over-cold bend sonar rays up instead of down and result in different appearing refraction bands (A). In this case the refraction is not as range limiting and bottom details (B) can be seen beyond the displayed refraction boundary.

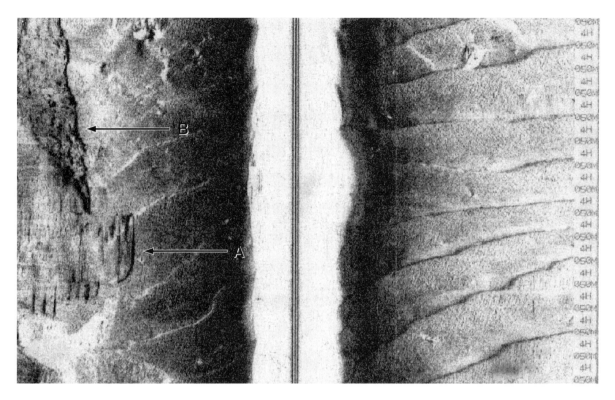

Record 108: Man-made thermal discontinuities are also detected by side scan sonar. At (A) two tug boats and a barge are moored. The center craft is discharging hot water and creating a thermal plume which can be seen dissipating down current (B).

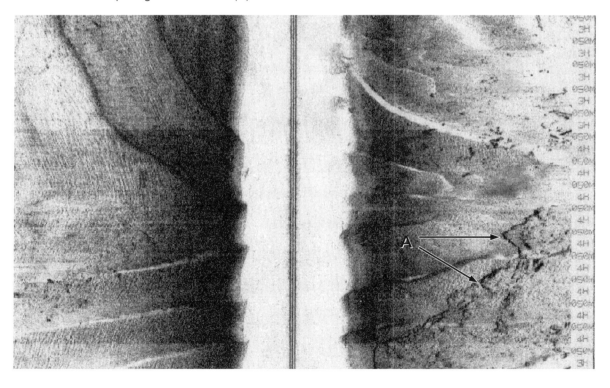

Record 109: The underside of ice provides enough back scatter to be detected by the sonar system (A). Although ice does not cast shadows on the seabed it might be mistaken for a mid water or bottom target.

Naturally occurring thermoclines are not the only density gradients found in the underwater environment. Salinity differences such as those found at outfalls and estuarine boundaries cause similar ray bending. Marked differences in temperature or density can act as a strong reflector as well as causing refraction. This effect in sonar data is often seen at hot or cold water discharge points. Imaging these discharges provides information about their source and dispersal.

In Record 108, the sonar was towed over an area of topography with large sand waves. On the port side, two tugboats and a barge are moored (A). The center craft is discharging hot water overboard (B). The difference in density between the hot water and the surrounding water provides a strong reflector and the sonar images the plume as it dissipates into the environment. The water, hottest at it's source, is a very strong reflector and it becomes less so as it mixes and dissipates.

When considering the effects of water density on side scan sonar systems, we must include ice. Armored cables should be used when operating in waters where ice might be encountered as the sheer weight of floating ice could severely damage light weight cables. Record 109 shows an area surveyed near ice covered water. The angularity of the underside of the ice flow (arrow) provides sufficient backscatter to make it easily recognizable to the operator. Although, due to the sonar geometry, the ice does not cast a shadow on the seabed, the feature could be mistaken for a midwater or bottom target if the operator was not aware of ice in the area.

OUT OF RANGE RETURNS

As in radar, the successive outgoing pulses of the sonar are carefully timed and synchronized with the display. As discussed in the previous section concerning noise, the outgoing pulses in sonar do not disappear at the end of the scanned range but rather continue out into the environment insonifying the seabed on the way. There are reflections from these pulses that return to the towfish but they are usually of low intensity. More importantly, the TVG circuitry is set low enough at the start of a scan that the returning echoes from beyond the set range do not interfere with the usual recording process of the sonar. Sometimes, if there are very strong reflectors out of the established recording range, their images appear on the record. These *out-of-range-returns* often appear from hard reflectors that are just out of range and thus they will be printed soon after the next outgoing pulse starts. In a corrected record these are not usually seen because they would occur in the non-displayed water column section of the record. Sometimes these returns are seen on the outer edges of the record where the gains are set high by the TVG circuitry. When they do occur further out on the record, they will have no shadow. The real location of a target causing an *out-of-range-return* is determined by increasing the range setting of the sonar.

Two examples of this phenomena are shown. In Record 110, on the port channel is an image of a target that is out of range of the sonar. The target is a seawall 65 meters from the towpath. Note that, although it is a straight line, and does not seem to be a natural feature, it casts no shadow.

Record 111 was made at a construction site where caissons were positioned on the seabed. One caisson is seen at the outer ranges of the 50 meter scale record (B). A large portion of the caisson is out of range. The structural characteristics of the sheet pilings, however, makes them very hard reflectors, sending back a strong signal to the sonar from beyond the printed range. This shows up on the record in the water column as a dotted line (A). Since the signal amplification is very low as this portion of the record is printed, the reflections are no where near as dark as the actual pilings, but they are clearly distinguishable.

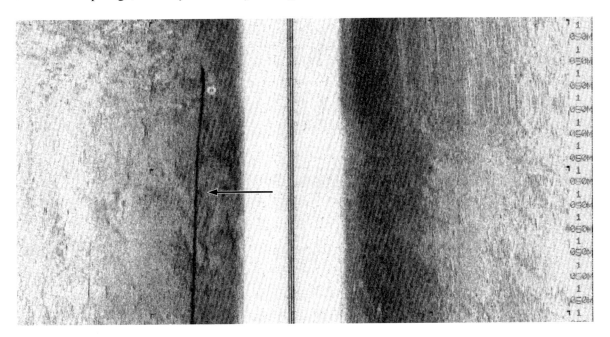

Record 110: Hard reflectors out of the set sonar range can be imaged by the sonar. These out-of-range returns can be seen in other scanning systems such as radar. Although they rarely occur in standard sonar operations, the out-of-range return will cast no shadow on the seabed. Here, an image of a sea wall (arrow) was generated by a structure approximately 65 meters to the port side of the towpath.

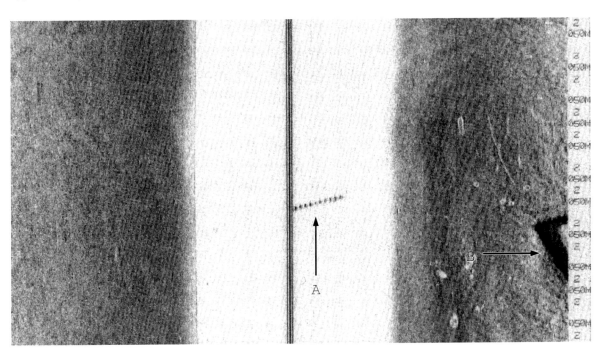

Record 111: Another example of out of range returns (A) is shown here generated by the hard edges of a portion of a cassion (B) which is out of the range set on the sonar recorder.

SURFACE REFLECTIONS

The vertical beam of the side scan is very wide. On its outward propagation, it will encounter the bottom, the surface, and anything in between that may be in the water column.

Although most sonar operations are tasked with imaging only those features on the bottom of the sea, the fact that we get return signals from the water column and the surface is very useful. The water column shown in the following records is in uncorrected records. Water column is also displayed in the third channel of the sonar system and can be viewed while generating fully-corrected sonar data.

The depth profile displayed in both the water column and the third data channel is not the true water depth, but the depth of the seabed below the towfish (see chapter 4). At the same time, the upper lobes of the sonar beam may reflect from the surface of the water and allow the system to display the sea surface directly above the towfish. In Record 112 the distance (A) is translated into the range from the fish to the surface (depth of the fish) while the distance (B) represents the height of the towfish. The sum of shortest range displayed to the first bottom return and the shortest range to the first surface return is equal to the water depth.

The relative positions of these returns will often provide the operator with valuable information about the vertical position of the fish. Knowing this position is important for gaining quality sonar images since the fish should be at the

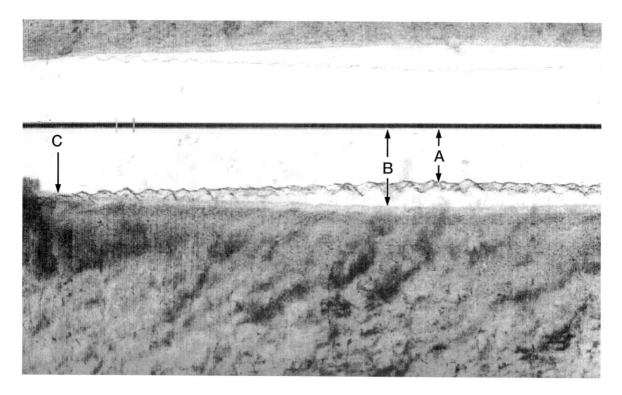

Record 112: The surface of the sea will act as a strong reflector to the sonar beam in many cases and will be imaged along with the seabed on the sonar display. The first surface return is a good indication of fish depth while the first bottom return represents fish height. When these two returns cross one another (C) the fish is midway between the surface and the bottom.

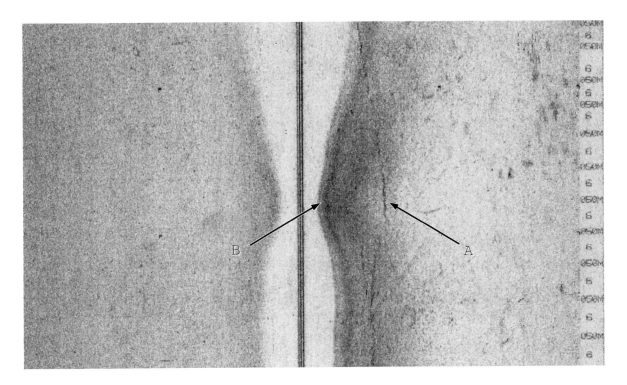

Record 113: The image of the first surface return moving to a greater range (A) while the first bottom return moves towards the trigger pulse (B) shows the towfish is being lowered rather than the water depth decreasing.

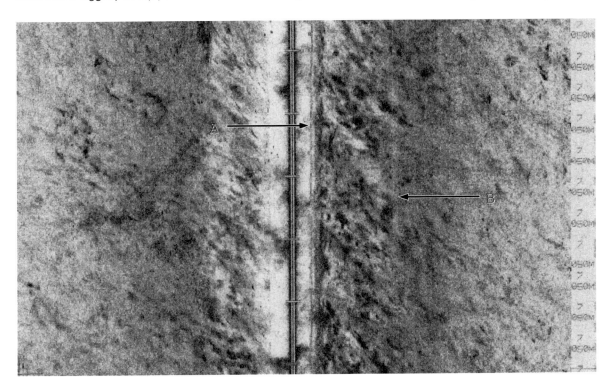

Record 114: When the sea surface is relatively calm, much of the incident energy from the sonar beam reflects out and away from the towfish. However, a rough sea surface will return some acoustic energy back to the towfish creating sea clutter across the displayed sonar image. These conditions provide a clearly delineated first surface return (A) but the resulting clutter can mask bottom targets and the first bottom return (B).

prescribed height above the bottom. For instance, when the first surface return and the first bottom return are superimposed upon one another (C in Record 112), the towfish is halfway between the bottom and the surface.

Accurate and timely interpretation of these returns are also helpful in understanding whether a decrease in fish height is due to decreasing water depth or the towfish assembly dropping. In Record 113 the water is not getting shallower; the fish has dropped deeper. The operator determines this by the fact that the first surface return has moved away from the trigger pulse on the record. When the towfish is towed near the surface, much of its transmitted acoustic energy will encounter the surface as it propagates away. If the water surface is relatively calm, this energy reflects and travels away from the towfish. If the surface is rough (from waves, for example), as in Record 114, some of this energy is reflected back to the towfish. Although these conditions will provide a strong first surface return (A), they sometimes mask midwater and bottom features including the first bottom return (B).

The fact that the sonar insonifies the surface, means that it registers even slight changes in the sea conditions. The light backscatter from the sea surface in Record 115 occurred after it started to rain on a smooth sea. To confirm that the rain was causing the effect seen in the record, the survey vessel was turned to starboard causing the towfish to bank (roll) slightly. As a result, the port transducer angled up, directing it's beam more towards the surface, while the starboard transducer was pointed downwards. As the record changed, the fact that the rain was causing the backscatter visible in the water column portion of the record, was confirmed.

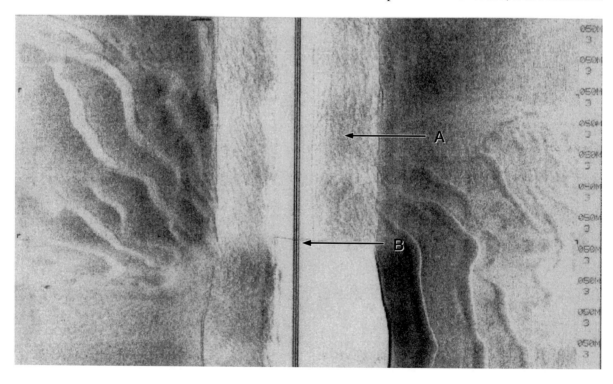

Record 115: Rain on a smooth sea creates surface roughness that may be detected by side scan sonar. Here, rain on the sea surface (A) was confirmed during a hard turn at (B) which caused the towfish to roll bringing the port transducer up and the starboard transducer down. This confirmed that the shading of the water column portion of the record was a result of increased sea surface roughness.

Sound reflected off the underside of the sea surface from seabed targets can also be received by the towfish. This gives false range information to the system. The effect is most often seen when survey operations are performed on smooth seas, in shallow water and where specular reflectors or very hard targets are found.

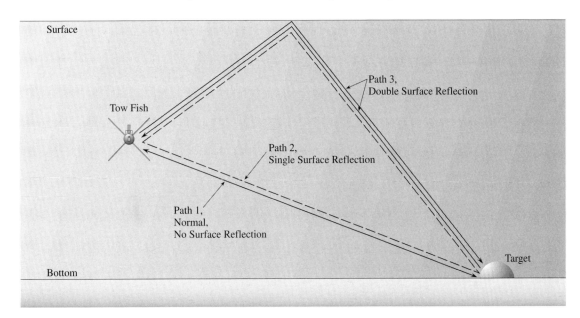

Illustration 116: In sonar geometry there are three paths by which the acoustic energy can travel from the towfish to the target and back. The primary path is directly from towfish to target in a straight line. When working in shallow water with calm sea states, acoustic energy can also reflect once or twice off the surface in its round trip travel. Using these three paths the sonar will display two or three images for one target (see Record 117).

Record 117: Generated in smooth sea conditions, this record contains an image of a steel shipwreck. The triple images of the starboard rail (arrow) are a result of multi path reflections of the sonar beam.

Three paths exist by which the output pulses of the sonar may insonify the target and be received by the towfish. Diagram 116 shows these paths. Typically, when this *multipath* phenomena occurs, sound received by the towfish comes by way of all three paths. A *specular reflector*, such as a cylinder, is similar to a hard target in that a large percent of incident energy is reflected back to it's source. The effect is seen with cylindrical navigational aids, pipelines and cylindrical mines or torpedoes.

Record 117, made in smooth sea conditions, shows a shipwreck in 10 meters of water. The ship is constructed of steel making it's components good sonar reflectors. The bow is evident with debris laying on the seabed on the port side of the wreck. The multipath effect is seen on the starboard rail of the ship and at the bow hold opening (A). The "ladder effect" that is seen in the seabed debris lying on the bottom (B) is not multipath in this example but rather reflection from frames attached to hull plates that fell away from the wreck.

This multipath effect was once attributed to a phenomenon called "Lloyds mirror bands". It was theorized that the surface reflected rays interfered with the direct sonar pulse and caused along-track banding on the sonar record. Much of the banding seen in records generated by the early sonar systems, and attributed to Lloyds Mirror Bands, was actually due to poor beam forming. However, occasionally multipath interference did occur when the correct sonar geometry was present.

WAKES

Wakes of objects passing through the water can be seen in sonar images. Sailboat, powerboat, and even rowboats leave discontinuities in the water that are detected by sonar. Wakes typically do not cast shadows (although they sometimes *quench* the sonar signal) and are often linear in overall appearance. These features help the operator to distinguish wakes, particularly older ones, from bottom features. Vessel wakes are very complex surface entities and they persist in the environment long after they become invisible to the eye. Sailboats add air to the water through waves caused by their bow and stern wakes. Powerboats, in addition to the bow and stern wakes, have a propeller that causes *cavitation* in the water which leaves large amounts of undissolved gas in the water. Outboard motors also churn the surface water and put exhaust gasses, that make good sonar reflectors, in the water. Powerboats dump hot water overboard that causes thermal discontinuities in the water as well. In quiet environments, rowboat oars cause enough disturbance to add small amounts of air that are evident on sonar. Since wakes can mask bottom targets, detrimentally affect a sonar signal, and are often overlaid on the record of the bottom, it is important to recognize them as an unnatural occurrence when they appear.

In Record 118, a side scan survey was performed in a busy harbor. Here, several slowly deteriorating vessel wakes, left by craft traveling approximately 60° to the survey vessel path, can be seen.

Normally, the wake of the survey vessel is not imaged by the sonar on straight runs. This is partially due to the downward tilt of the transducers in the towfish. However, in small area surveys where the towing vessel is navigating lanes and there is sufficient overlap to each lane, the survey vessels wake may be seen on

Record 118: This record was generated in a busy harbor way. Disintegrating powerboat wakes are visible crossing the survey track at an oblique angle. Gases generated in the water from both cavitation and air mixing by powerboat propellers provide enough discontinuities in the water column to be imaged by sonar.

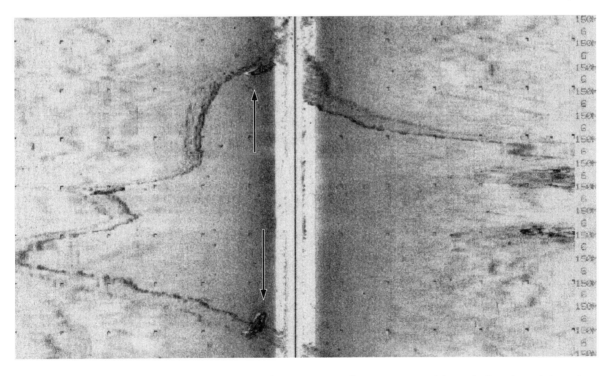

Record 119: The survey vessel's own wake can even become a problem during target imaging operations. In the record above, multiple passes made by the survey vessel near a small shipwreck (arrow) left a tell-tale wake. The image of the wake is overlaid on the image of the seabed.

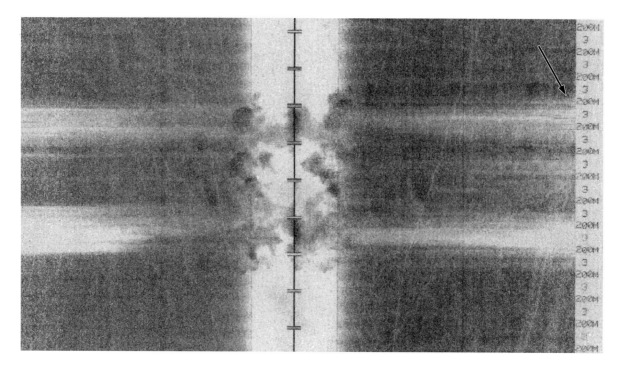

Record 120: In the record above, the sonar was towed through the wake of a ocean going tug and barge. The bow wake of the barge was wide enough to appear as two separate wakes. There is enough air in the water from the wake to partially quench the sonar signal at the outer range (arrow).

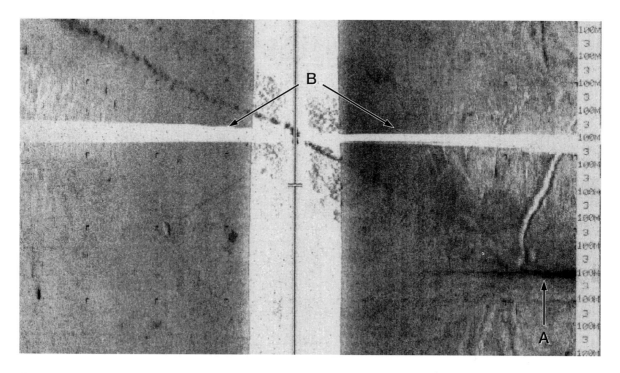

Record 121: Towing through a recent wake can cause severe quenching of the sonar. In the record above, a powerboat overtook the survey vessel on the starboard side at position (A). The other vessels propeller noise is evident here. The boat then passed diagonally in front of the survey vessel. When the towfish encountered the other vessels wake, the sonar signal was completely quenched (B).

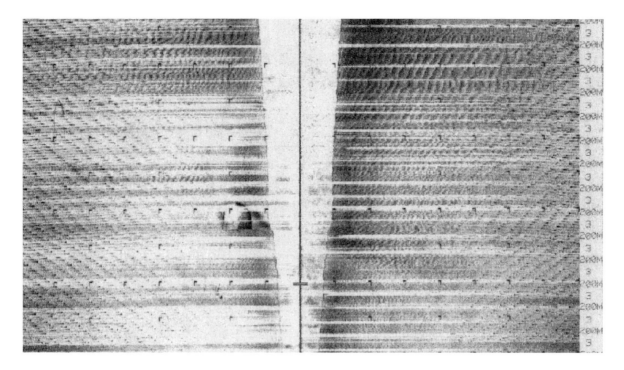

Record 122: When working in shallow water the towfish must be towed close to the surface and may actually be in the survey vessels own wake. This can cause sonar quenching. In the record above an oil tanker sunk on a shoal in 1976 was imaged with the towfish in the wake of the survey vessel. Since the towfish was deployed on the starboard side quenching was most severe on the port channel.

Record 123: Here, a coastal waterway was surveyed with the towfish deployed off the starboard side of the survey vessel and very close to the wake. The shoreline can be seen the starboard channel (A) and quenching can be seen on the port channel (B). The survey vessels own propeller created acoustic interference (C) until the fish was lowered slightly.

each successive pass. In these cases, it is helpful to plan the survey such that the lanes work successively into the tide and/or wind so that these forces tend to push the wake away from the next lane.

The wake of the survey vessel itself is sometimes a problem during target imaging. In Record 119, the survey vessel made several passes on a small shipwreck, each time leaving a telltale wake near the target. The two bottom images of a target in this record are of the same shipwreck.

Wakes also contribute enough discontinuities into the water column to cause quenching or blanking of the sonar signal. If the sonar is towed close to the surface, or the disturbance of the wake is massive enough to disturb a large vertical portion of the water column, it will affect the sonar data. In Record 120, the sonar was towed through the deteriorating wake of an ocean going tugboat and barge. It looks like two separate wakes, but this effect is from opposite sides of a large bow wake and there is enough air in the water column to partially *blank* the sonar signal at the outer ranges. In Record 121, a power boat overtook the survey vessel on the starboard side (propeller noise is visible on the far right channel) and passed in front of it. As the survey vessel towed the sonar fish through the wake of the power boat, the sonar signal was completely quenched leaving a white streak across the record at that point. The signal blocking is very thorough here. The power boat left a large amount of gas in the water and the towfish was towed near the surface directly through the wake.

Other vessels' wakes are not the only cause of quenching. When towing in very shallow water, the operator may want to raise the towfish high in the water column in order to avoid collisions with the bottom. The primary problem in this case is that the towfish may continually work in the wake of the towing vessel and this can cause quenching for long periods of the time (See Record 122). Although the water became deeper as the track progressed, the operator did not lower the fish since they were to return into shallow water. In a situation where the water is very shallow, one should tow from the bow or use other methods to keep the towfish out of the survey vessels wake. In Record 122, the fish was towed slightly to starboard, making the quenching more severe on the port channel. This quenching is more noticeable in the longer ranges. In Record 123, the survey vessel was operating at an increased speed. This had the effect of bringing the towfish high in the water column and partially into the vessels wake. The towfish was towed from the starboard stern on a short stay. Quenching is visible at the outer ranges on the port channel along with some evidence of the vessels propeller noise. The light area 60 meters to starboard is the shoreline. Slowing the ship down reduced the acoustic level of the propeller noise, but since the towfish was still operating in the vessels wake, quenching was not reduced. Another example of quenching and propeller noise is seen in Record 124. On this survey the operator was running at 200 meters per side, using a tow speed that brought the towfish high into the vessel wake. On the lower port channel both quenching and propeller noise is evident. Slowing the survey vessel eliminated these effects by allowing the towfish to sink away from the wake and the vessel.

Wakes also affect data when generating slant range corrected sonar records. The sonar uses the range to the bottom to calculate correction factors and to remove the water column from the sonar record. This range figure is taken from one of the

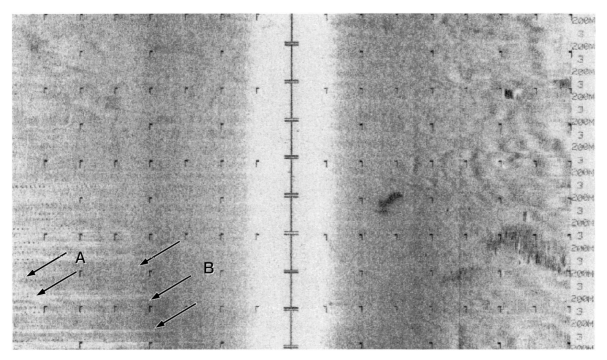

Record 124: At long ranges, sonar signal quenching occurs more easily since signals returning from great distances are significantly weaker than those from the near ranges. In the record above, propeller noise from the survey vessel can be seen at the outer ranges (A). Even with these high gains wake quenching occurs (B) because the towfish is close to the wake.

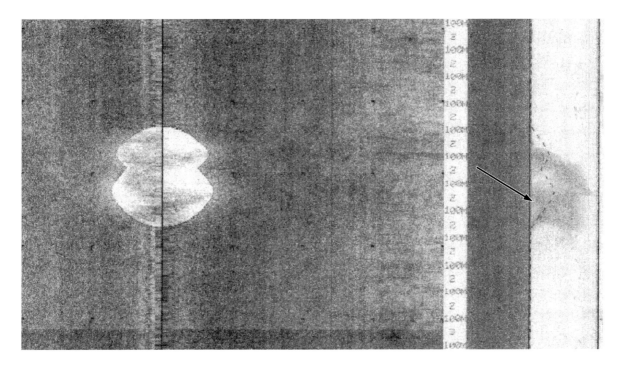

Record 125 When generating corrected records side scan sonar altitude detection may "lock onto" water borne discontinuities and display false altitude information. In the above record, the sonar was towed through the wake of an ocean going vessel. The sonar temporarily lost bottom lock as can be seen in the third data channel (arrow).

side scan channels. If there are major discontinuities in the water column, the bottom tracking electronics may "lock onto" these discontinuities instead of the bottom and display improperly corrected data. Because the gasses in wakes are a strong reflector, encountering them can sometimes result in the system losing *bottom lock*. Record 125 shows the result of the survey vessel passing through another vessels wake (evident in the data channel (arrow)) and the bottom tracking indicator sensing this as a false bottom. If the data is stored on tape, it should be rerun with the recorder set to manual correction for this data section. If the data is not being stored, the operator should realize that when towing through other vessel wakes, the system should be set to the manual mode.

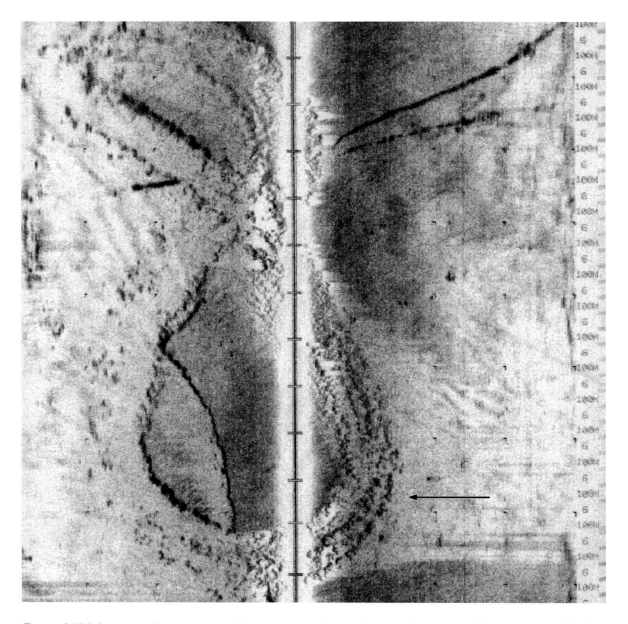

Record 126: Large scale sonar quenching can occur in very busy water ways and become range limiting for survey operations. In the record above, many vessels were traversing the area to be surveyed with one power craft creating a figure-eight wake in front of the survey vessel. Note the overall reduction in return signal strength beyond these wakes (arrow).

Wakes are not a problem for operations in deep water when the towfish is towed well below the surface, but they seriously affect operations where the towfish must be towed near the surface. The combination of signal quenching and target masking in a busy waterway may cause the cessation of a planned operation. Record 126 was made in a harbor where many pleasure craft were traversing through the survey area for several hours before the sonar operation started. During the survey the fish was towed near the surface due to the fluctuating bathymetry. As a result of the disturbed surface water, gains had to be turned higher than normal and in the "figure eight" wake, that was generated by a small outboard motor, the signal level returning from the outer ranges was significantly reduced.

BIOLOGICAL TARGETS

In most fresh water and marine environments, sonar surveys may detect the presence of individual and schools of fish. Although very large aggregations and high population densities will actually mask small bottom targets, traces of fish on sonar records are not usually bothersome if they are immediately recognized as biological targets. Fish act as sonar reflectors both by the gas carried in their air bladders and the makeup of their bodies as discontinuities in the water. In medium range applications, schools of fish might be indistinguishable from the bottom backscatter, however, their shadows are cast on the record if the sonar geometry is appropriate. When closer to the towfish, fish are imaged individually if they are large enough and the sonar range setting is low.

In Record 127, the towfish was towed through a school of menhaden (*Brevoortia tyrannus*). The animals are so close to the towfish and the range setting is low enough that the individual fish can be resolved. The sonar geometry does not allow the system to display all of the shadow that would be cast by the fish if the range were set higher. Usually this type of target in the water column does not seriously affect either general or small target search operations.

In Record 128, there was a small number of bluefish (*Pomatomus saltatrix*) swimming near the bottom in the same direction as that of the tow. This had the effect of elongating the images of each individual fish. Since they were relatively close to the bottom, their associated shadows are well within the sonar range setting. The association of shadows with targets makes a sonar record easier to interpret.

The dark areas on Record 7, at the end of Chapter III., are schools of herring (*Clupea harengus*) aggregated during the winter time. The shadows cast by some of the schools indicate their vertical distribution. During the warmer months of the year this species does not school heavily and are individually small enough (approximately 15 cm in length) to be indistinguishable from bottom backscatter.

Record 129 shows another school of menhaden that are approximately 20 cm in length and densely schooled together to form a significant acoustic reflector. The range setting in this record is 100 meters as opposed to the 75 meter setting on Record 127. Here we begin to loose the sense of the individual fish and see the sonar target as an amorphous grouping. This record is also a good example of a biological target's shadow providing schooling distribution data. The shadow and its relationship to the target, indicates that the school is close to the seabed and hemispherically shaped at the base.

When running sonar operations in the uncorrected mode, fish in the water column are easily identified as in Record 130 where the sonar was towed through a school of mackerel (*Scomber scombrus*) about 35-45 cm in length, swimming around a small shipwreck. These fish were not densely packed, but they were big enough, and the sonar range was low enough, for the sonar to display the individual fish. The fish are swimming high off the bottom and their shadows are not clearly evident. The record shows a great deal of sea clutter indicating rough surface conditions when the record was generated.

When operating in the corrected mode, shadows cast on the seafloor without apparent associated targets, as in Record 131 may be fish in the water column. In this particular case the sonar was towed above a school of bluefish. If the fish are large enough, their existence may be confirmed by examination of the third channel of the sonar record.

Animals that use sonar are sometimes attracted to the towfish. This is troublesome if the animals are persistent and interrupt the survey process by swimming near the transducers and blocking the signal. Three records were made during one survey where porpoise (species unknown) swam alongside and underneath the sonar towfish. This is a case where animals in the water column, if large enough, fully mask the signal. In Record 132, the animals were almost in contact with the towfish. The black dots on the record are the porpoises sonar being registered by the side scan. Viewing sonar data in an uncorrected mode would provide the operator with a higher quality image of objects in the water column than that displayed in the third channel. In Record 133 the porpoise continue to cause noise on the record and cast shadows on the seabed. Here, in the uncorrected mode, the returns from the animals are seen near the trigger pulse as they swim under the towfish. In addition to causing noise on the sonar they also cast significant shadows on the bottom. In Record 134 the range setting has been dropped to get an enlarged image of the porpoise swimming below the towfish. The image shows the up and down swimming motion characteristic of the animal. Of all the animals that might be encountered during sonar operations, porpoise are particularly curious of the towfish and its acoustic output.

Most underwater plant life does not provide significant discontinuities that can be recognized on a side scan display. Other forms of marine life provide varying levels of backscatter when compared with that of the surrounding seabed. In Record 135, a sandy seabed with large and small sand waves supports large shellfish beds which are seen as dark areas in the sonar data. Side scan sonar has proven to be a commercially viable tool for locating shellfish resources such as scallops and mussels.

With all of the different factors that can influence side scan sonar results, it may seem that record interpretation is a difficult process. In reality, interpretation of most sonar records, although qualitative, can be a straight forward process. Only occasionally do sonar records have unpredicted results. Seeing and understanding different phenomena in sonar records will give the operator a broad base of knowledge important for accurately interpreting unpredicted results.

In general, making a record interpretable means applying the equipment so that the sonar data shows the surveyor what he needs to see. An interpretable record for a survey of a soft mud seabed would be tuned with high gains but still be light in appearance. The records generated during a survey of hard sediments would be tuned and interpreted differently as would records made during a survey for very old shipwrecks or modern man made debris fields. Above all, it is important for the sonar interpreter to remember that the application of the system and the purpose of the survey will dictate the methods used to generate easily interpretable records.

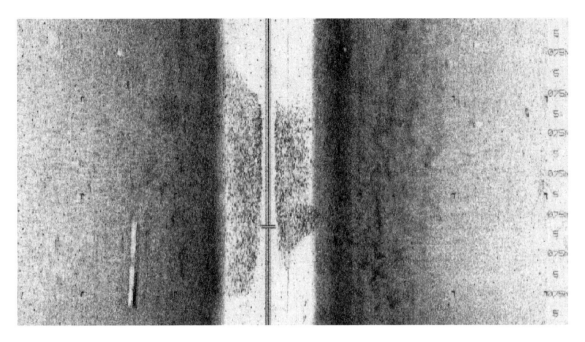

Record 127: Biological targets are often imaged by side scan sonar. In the record above the sonar was towed below a school of small fish.

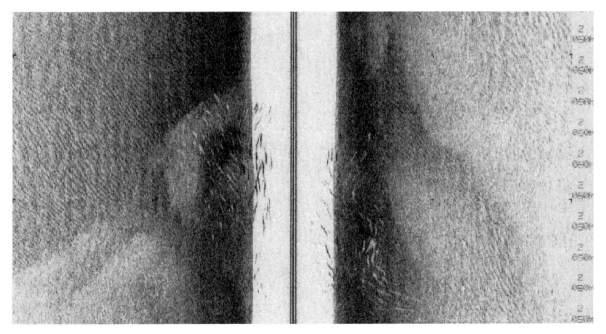

Record 128: Large fish can be individually imaged by sonar. This is particularly true if the fish are swimming in the direction of tow as in the record above.

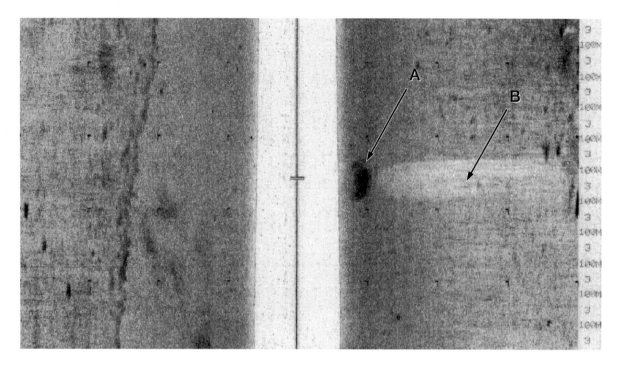

Record 129: When the sonar is set to long ranges, schools of fish may appear as single masses on the sonar record. A school of fish in midwater (A) will l cast a shadow (B) separated from the school's sonar return.

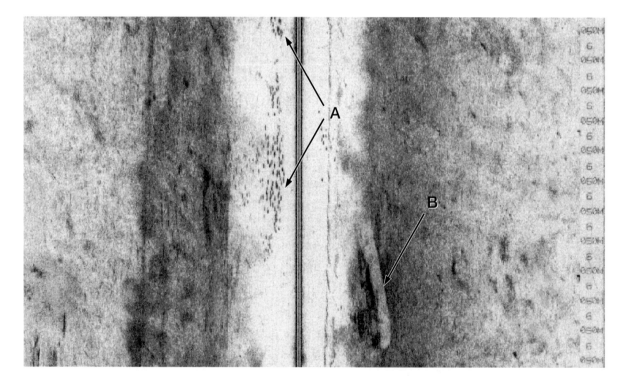

Record 130: shadows of individual fish will be displayed on the sonar record if the sonar geometry and operating environment is correct. In the record above, the sonar was towed through a school of fish (A) swimming around a small shipwreck (B). The fish shadows are barely distinguishable due to severe sea clutter overlaid on the seabed image.

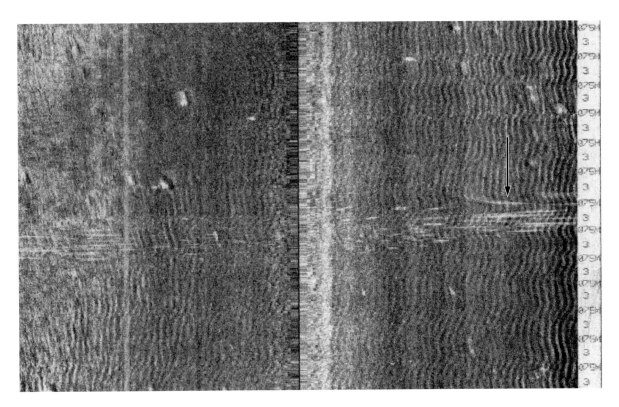

Record 131: When generating corrected records with the water column removed, fish in the water column will not be displayed but their associated shadows (arrow) may be seen in the record.

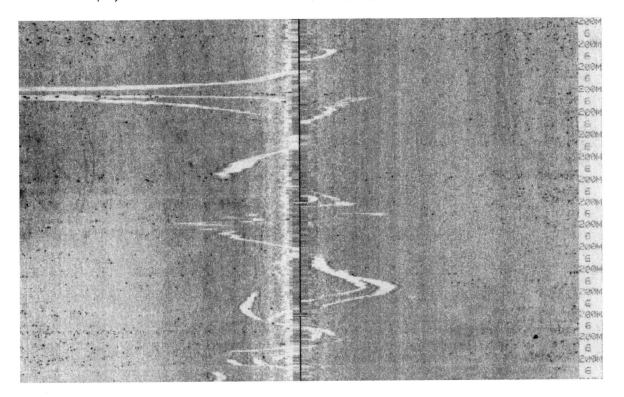

Record 132: Large animals in the water column will cast significant shadows on the seabed. In the record above, the shadows are accompanied by acoustic interference. See records 133 and 134.

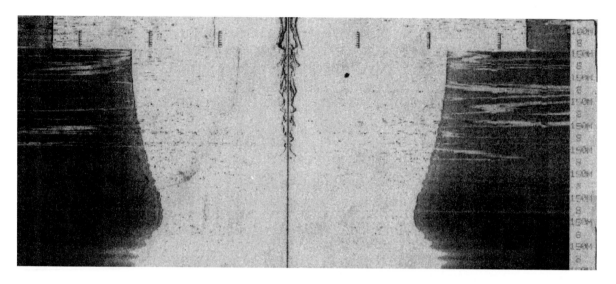

Record 133: After generating record 132 the operator shifted to the uncorrected mode in order to examine the water column region of the record more closely. Here he could see sonar reflectors near the towfish transducer (arrow).

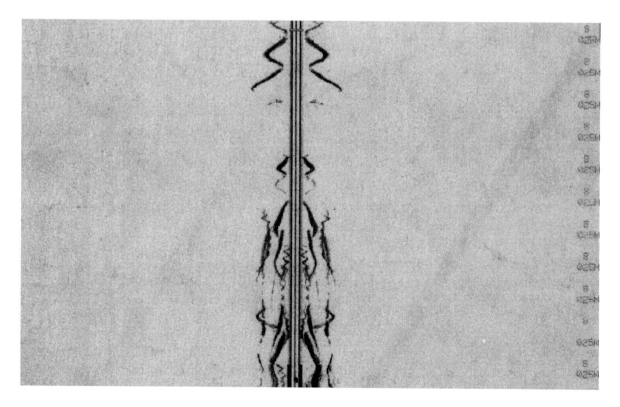

Record 134: After generating record 133 the operator reduced the sonar range to 25 meters. This provided a larger image of the sonar reflectors near the towfish. The body undulations of porpoise swimming in the direction of tow and just under the towfish can be seen in this record.

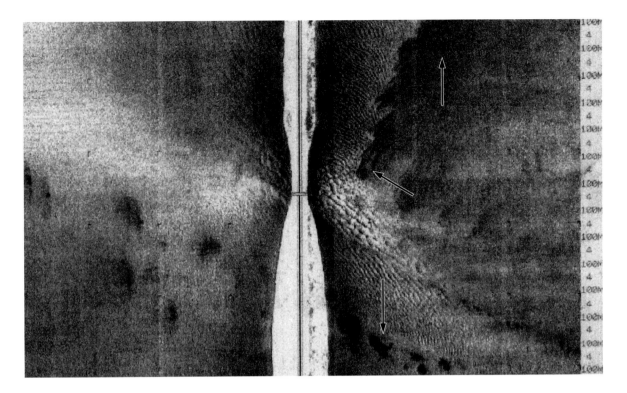

Record 135: Because bottom roughness is changed by large colonies of shellfish, these areas can be delineated by sonar. In the above record, the dark areas on the sonar record (arrow) are shellfish beds.

Chapter 9:
Mosaics

Illustration 136: This side scan sonar mosaic is a portion of a two-square-mile image of the wreck of the *USS Monitor* and the surrounding seabed. It shows a deep scour stretching down current from the wreck. *Courtesy EG&G*

Efforts have been made to improve the quality of underwater imaging ever since the earliest experiments in underwater photography. Acoustic imaging gained ground during the years of World War II. Since that time there have been essentially two types of imaging systems that could be used in underwater applications: optical and acoustic. Only a few decades ago, there was a large gap between these two technologies. Using acoustic imaging, the scientist or surveyor could have reasonably large area coverage but the resolution was poor. Optically, early still cameras could obtain high resolution but coverage was limited to only a few meters. Since that time, underwater imaging technology has been filling the gap steadily with a number of notable milestones. In optics, the development of instrumentation like the Electronic Still Camera™ (Marine Imaging Systems) has markedly increased the coverage of the optical system; while in acoustics, the application of 500 kHz side scan sonar has brought almost photographic resolution.

Refining this examination to sonar alone, there still exists a gap between resolution and coverage. Systems have been developed to image large sections of seabed in one swath with the resulting data on one record. These ranges have exceeded 5 kilometers per swath but the resolving power of these systems is rarely more than several meters. Conversely, the 500 kHz side scan resolves sub-decimeter sized objects but its range seldom surpasses 300 meters in swath width.

One solution that fills the gap between resolution and coverage, thus providing large scale, high resolution images, is the production of mosaics or paste-ups of multiple records from adjacent survey lanes. This entails running a series of adjacent lanes and matching each record to the next to produce a high resolution, and large scale image of the area surveyed.

Mosaics are valuable because it is sometimes necessary to understand what a large area of seafloor looks like in order to interpret data from side scan sonar. Further, it is not always possible to look at each strip of data and come to an accurate conclusion. In addition, if these mosaics are generated as a 1:1 scaler plan view of the seabed, features can be scaled directly off the records and their true size and position could be easily determined.

Side scan mosaics were difficult to make in the 1960's and 70's. Early analog sonar systems were designed to generate only one record and had no means to store the data for the user to make duplicates. If the original record was lost or damaged in the process of pasting, that lane would have to be run again at sea. Furthermore, the side scan could only produce uncorrected data that contained the water column as well as the associated slant range and speed distortions.

One of the most delicate controls required in matching sonar data from adjacent lanes is that of mating the sonar chart speed with that of the survey vessel. With the early systems, very small speed differences between one lane and the next resulted in two records of different lengths that could not be matched. Numerous efforts at creating mosaics during these years were tried with some limited success and only after great effort.

With the introduction of the digital sonar systems and multistyli or thermal recorders in the 1980's, the operator can correct side scan data for vessel speed and slant range, remove the water column, and digitally store data which permits further amplitude adjustments and record enhancements. Side scan mosaics are then created by piecing together adjacent records with sufficient overlap to match identical features and navigation shot points. A true plan view map of the seabed is created this way and the original or recorded side scan data used as a base map.

Side scan mosaics have become a popular data presentation in the oil industry for block clearance, and for pipeline and geological mapping surveys at drilling sites. Side scan mosaics are also used for many applications besides geological surveys. Mosaics have been prepared to aid in planning benthic monitoring programs, dredge spoil monitoring programs for the Army Corps of Engineers, fisheries applications such as fish and marine life habitat surveys, pier and dock site-assessment surveys, and marine archaeological surveys.

EQUIPMENT

In addition to the standard digital side scan equipment, one of the most important pieces of equipment for creating mosaics is a digital data recorder that will store all of the sonar data gathered and allow the user to rerun (and reprocess if desired) the data after at-sea operations are complete. This lets the surveyor make duplicate records in the event of damage to the originals, and to match the vessel speed accurately on different and adjacent lanes when duplicating the record.

Navigation is more important during mosaic data gathering than most normal sonar survey operations. Tracklines must be run very precisely to avoid creating too much, or not enough, overlap. Speed data also has to be provided from a navigation system because consistent tow speeds are important for generating the final image for the mosaic creation.

DATA GATHERING TECHNIQUES

In most mosaics, the records generated at sea are not used for the final image construction. The raw data gathered at sea must be recorded. Hard copy records or video images are used during the survey but the recorded data is what is used in the laboratory for final image construction.

For mosaics, obtaining high quality, uniform, side scan data is imperative. Data should be collected during calm sea periods to minimize distortions from towfish instability. The towfish should be maintained at constant height above the bottom, and a steady OTG vessel speed should be maintained. The towfish layback distance from the navigation antennae should be carefully monitored and recorded during operations.

Precision navigation systems should be used during the survey. These should provide a plus or minus one meter accuracy, one second update rate, a steering display for the helmsman, and a speed output to the side scan system. Side scan lines should be run as straight as possible and line spacings selected to provide 25-50% overlap of side scan data. Event marks put on the record and stored with the sonar data are very important during both data gathering and post processing. It is very useful to have event marks equal to (or a multiple of) the scale setting.

For example, if the system is set to the 50 meter range and events are 50 meters apart, on examination of the record, the width of one channel should be equal to the length between events. When on 100 meters with events spaced at 50 meters, two of the events should equal the width on one channel.

It is useful to run some test lines over the area to be surveyed to determine the best gain setting for the final mosaic. Once begun, mosaic data gathering should proceed in an orderly fashion. The survey should proceed from one lane to the next because, if the work cannot be completed in one day, there may be differences in data quality from one day to the next due to weather conditions or water quality. Lighter sediments often appear nearly pure white in the sonar record, but this is improved by later photographic reproduction. Records in mosaics that are too dark may hide details.

When a survey is begun, the operator should keep in mind that when this data is combined to form a mosaic, the lines will be straight, even if the vessel track was not. It is therefore, more important to keep the vessel on a steady course, instead of continually adjusting for errors in navigation. If this is not possible because of pre-plots (plotted tracklines that must be followed) the vessel captain should make very slow changes to the vessel track. Instead of oversteering each side of the trackline, as might be done in other types of sonar operations, straight line corrections provide better overall data for the final mosaic. When surveying, an effort should be made to maintain the fish height at roughly the same distance off the bottom. This is not always practical but the side scan will compensate for this error up to 47% of full range.

DATA PROCESSING

The first step in constructing the actual graphic mosaic, is to produce a navigation post plot at the same scale as the side scan data. For example, if, on the display of the sonar system, 10 cm equals 100 meters for data collected at 100 meter range scale, then a navigation post plot at a scale of 1-1000 would be required. This would require a large flatbed or roller plotter and the final post plot would be pieced together into sections.

The second step in producing an accurate mosaic is to calculate the actual speed over the ground between navigation fix marks and replay a corrected graphic record with the new speed information. This is an extremely time consuming process, as one day of data collected in the field will require one day in the laboratory applying the speed corrections. If a reliable speed interface is available from a precision navigation system, or if data is collected from an area of low current velocities, speed correcting the side scan data may not be required.

Once the final speed corrected copy of the side scan record has been generated, the hard copy record is laminated with a thin plastic film using a standard 12 inch lamination machine. This will allow the records to be annotated and taped together without tearing or marking the data. Laminating records from thermal recorders is not necessary.

The final step in producing the mosaic is to align the side scan records with the navigation tracklines, matching navigation fix marks on the post plot with the records. This will require that the side scan data be offset to allow for the layback of the towfish. With short cable lengths in shallow water, the layback is calculated by monitoring the amount of cable deployed and knowing the towfish height. In addition to matching the center lines of the records and the navigation fix marks, identical geologic features on adjacent records will be aligned and the mosaic assembled into its final form by taping the records together on the back side using clear plastic tape.

Mosaics are most often presented in the form of a single photograph and it is important to determine the largest size artwork that is handled at the facility assigned to photograph the assembled mosaic.

SOURCES OF ERROR

The major difficulty in aligning side scan lines to form a mosaic is related to an improper vessel speed input which either compresses or elongates bottom features. Although the stored data can be replayed and speed corrections applied at the fix mark intervals, this results in an average speed over an interval, and does not correct for vessel speed fluctuations between fix marks. In addition, as tracklines are never perfectly straight, the amount of overlap between adjacent side scan lines varies and assembling the mosaic often becomes a best-fit situation. Rarely are side scan records cut into sections to exactly follow the tracklines because it usually provides a poor presentation of the data. Typically a compromise is made to achieve both accurate presentation of the data and an esthetically pleasing final product.

Another potential source of error in assembling a mosaic is the improper calculation, or estimate, of towfish position in relation to the ship's navigation antennae. This will occur when operating in deeper water with longer cable lengths. When using a long length of cable, the wire angle of the cable and the catenary is largely dependent on vessel speed and current velocity; thus, the towfish layback is continually changing. In addition, with multiple changes in the towfish heading, the towfish may not be located directly behind the vessel. In this case the actual towfish position can only be estimated by using conventional methods unless an acoustic tracking system is used.

Another source of error which makes the assembly of side scan mosaics and scaling of bottom targets difficult, is related to errors in slant range corrections on sea floors of varying slope and topography such as rock ledges or sand waves. As mentioned in earlier chapters, the slant range correction is based on the initial return from one of the transducers. In hard bottom areas of uniform depth, the system tracks the bottom well and spatially corrected data is produced. However, under certain conditions, such as turbid water, rough sea conditions, or shallow water, tracking the bottom is often difficult and slant range corrected data is hard to obtain.

Examples of side scan mosaics demonstrating different applications are presented in the following illustrations.

Illustration 136 portrays a portion of a two-square-mile large scale mosaic of the NOAA marine sanctuary containing the wreckage of the *USS Monitor*. The mosaic reveals that the remains of the *Monitor* rest on a sandy plateau over one-half-mile from a section of rugged hard bottom. One of the most striking features of the mosaic is that it shows a severe scour or bottom disturbance immediately down current of the wreckage. This scour extends for hundreds of yards away from the wreck. The construction of the mosaic demonstrates that this scour is not prevalent in other areas on the plateau and is most likely caused by the presence of the iron wreck.

This and another 500 kHz side scan mosaic of the *Monitor* site will eventually hang in the *Monitor* Museum. They also served in the planning of ROV operations to provide complete photographic coverage of the wrecksite to produce a photographic mosaic.

Another example of a side scan mosaic is in Illustration 137. This mosaic is a small portion of a larger mosaic which represents eight lines of data collected at the 100 meter scale, and provides bottom coverage of the proposed sewer diffuser site in outer Boston Harbor. The mosaic reveals the varied bottom topography at the diffuser site, including a steep rock ledge, bolder fields, and areas of fine grained sand. An overlay showing the diffuser alignment and the test-boring locations was also prepared. This side scan mosaic will aid engineers in the drilling of vertical shafts at the diffuser site.

A dramatic example of a side scan mosaic, which resembles an aerial photograph is Illustration 138. This side scan data was collected at the former town of Enfield, Massachusetts, which is now submerged under a large reservoir. As shown in this photograph, building structures and trees were removed prior to flooding the land back in the early 1900's. This side scan mosaic was produced as a demonstration project for the Massachusetts Metropolitan District Commission.

This mosaic was constructed from data that was not recorded on tape and speed corrected. Thus, features from adjacent records do not match exactly. However, the mosaic clearly reveals the old road beds, fields, major intersections, house foundations, and a disturbed area which was identified as the former cemetery after graves had been removed. These features match well with town maps of the region. More extensive side scan mosaics of former towns under reservoirs supplying municipal water are being planned by municipal water departments around the country. These mosaics will help identify the locations of former industrial properties under the reservoirs which may represent potential sources for heavy metal pollutants.

Mosaics can be constructed not only of large seabed areas many kilometers in width but also of small areas where high resolution is required beyond the one pass swath width of the sonar. Illustration 139 is a mosaic constructed from just a few passes in a small harbor in a pre-dredge assessment. The floor of the harbor is soft mud with few obstructions. The mosaic shows the relation of the pier face on one side to anomalies on the floor of the harbor and a dredge cut at the far right.

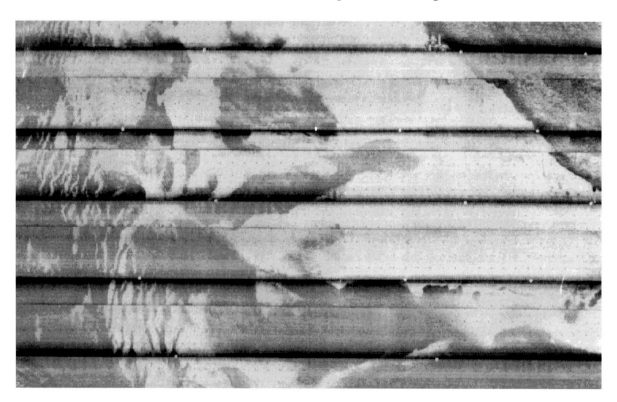

Illustration 137: The mosaic above is a portion of a survey performed at the proposed site of a sewer diffusing network in Boston Harbor, Massachusetts. It shows soft sediments and how they geographically relate to neighboring hard sediments and rock ledge. *Courtesy EG&G WASC*

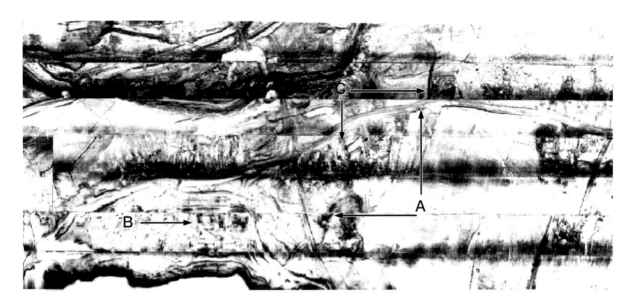

Illustration 138: This unique mosaic represents the bottom of a man-made reservoir. Here, a valley was water filled and covered a small town. Many of the buildings were removed and bodies were carefully removed from the town cemetery prior to the creation of the reservoir. However, the town streets are still evident (A) along with the cemetery (B) and building sites (C) of the now submerged town. Because the reservoir supplies municipal water, this mosaic helped identify potential sources of heavy metals. *Courtesy EG&G WASC*

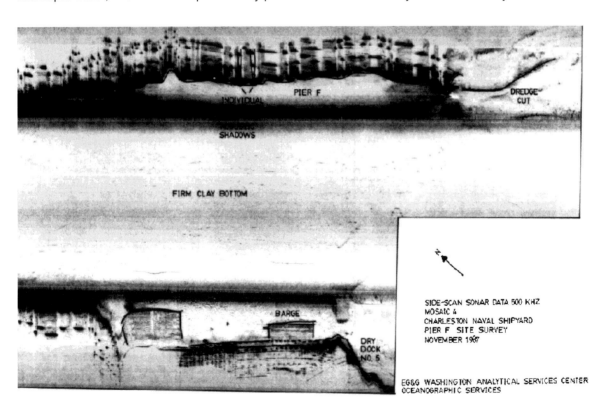

Illustration 139: Small scale mosaics also can be valuable for area assessment. The mosaic above was constructed from just a few short, high resolution runs in a Charleston, N.C. harbor. The mosaic shows the proper perspective and configuration of the harbor floor as it relates to piers. A scour from a dredge cutter head is in the upper right of the illustration. *Courtesy EG&G WASC*

POSITIONING BOTTOM FEATURES

Overall, the side scan mosaic is a useful tool in the interpretation of bottom conditions and sediment types as well as for search operations. However, the side scan mosaic is intended to give the interpreter a general overview of a specific area. When precise locations of an object or target are desired, the interpreter will have to use the individual lines in which the target exists and offset from the closest event. It is possible to overlay navigation data, but due to boat movement and speed inconsistencies, the mosaic should not be relied on to give exact locations of objects from the mosaic itself.

COMPUTER GENERATED MOSAICS

Computer based mapping systems are used by the research community, oil industry, and the military for large scale applications, and will become more widely used in coastal waters as these systems become more economical. They consist of sophisticated digital recording and processing modules which greatly reduce the time required in the assembly of side scan mosaics and minimize positioning and correctional errors. Two requirements of these systems are: 1. the use of an acoustic tracking system which provides an accurate and quick update of towfish position, and 2. greater storage capacity on optical discs and other mass digital storage mediums, which permits storage of time, vessel speed and navigation positions in a header with every scan of the system.

With these heavily based computer systems, the graphic records are not cut and pasted to produce mosaics, as these tasks are now handled by the computer. Recorded data can be quickly processed and accurate side scan mosaics are produced on board ship which aids in the planning of sampling and measurement programs.

An important development, as an outgrowth of computerized construction of side scan sonar mosaics, is the generation of bathymetric information and use of high resolution three-dimensional video displays. The technology of displaying side scan sonar data on video is not new; it was developed in the 1970's and early 1980's. In the 1990's however, far more advanced display systems are being employed in the high-end, complex sonar systems such as the three-dimensional Bathimagry™ (Triton Technology Inc.) system. These sonar displays give the sonar operator a three-dimensional image that shows the same amount of data visible on a two-dimensional paper recorder. Since the sonar data's bathymetric, or z-component, is known, the data is presented in a "vanishing point" display and the operator flies his point of perspective over the bottom terrain as he chooses. New data is generated at the "horizon" of the display and approaches the operator's viewpoint with each new outgoing pulse. With these systems, the operator has the same view as if he were being towed above the bottom and behind the towfish. Although these systems are expensive, the three-dimensional view makes small target recognition much faster and easier than conventional sonar systems. It is particularly valuable for wide swath, medium target and narrow swath, small target search operations. The bathymetric data generated from these systems is valuable for geophysical and geological surveys, and important to the marine scientist.

Chapter 10:
Detail Mapping

Illustration 140: The fishing vessel *Margaret Rose* struck the beach in a storm in 1962. This was one of the last occasions where the US Coast Guard used the Breeches Buoy to rescue shipwrecked sailors. The vessel was successfully salvaged and continued to work as a commercial fishing craft for many years before being wrecked in deeper water (see Record 141). *From the collection of William P. Quinn.*

Side scan sonar is often used by salvage and insurance survey specialists not only to search for targets but also to map large targets in detail prior to underwater salvage operations. When using the right procedures with side scan sonar, a shipwreck can be imaged from bow to stern and show each hold opening and any structural damage that has occurred. These procedures can be critical to an effective operation in that they identify components or details of the wreck that will allow salvage survey management to guide divers and remotely operated vehicles to the appropriate places on the wreck once intervention operations commence.

If the operation involves surveying a modern vessel to locate and optically inspect certain areas of a wreck such as below the water line or inside the bridge area, a detailed initial sonar image of the site will help the surveyor direct cameras to the proper place. For archaeological work on shipwrecks, a high resolution acoustic image will assist in determining the location of site components as well as the size and distribution of any debris fields. This determination can be made during planning stages of an underwater survey. For major salvage operations of a modern vessel, this process assists the identification of prominent components of the wreck before salvage operations start.

Known in the sonar field as "site component imaging" or, more properly, *detail mapping*, these operations are performed differently from those involving large area searches. Operations undertaken to investigate extensive areas for targets typically utilize long ranges, relatively high towfish altitudes, low sonar frequencies and a variety of vessel speeds. The operators involved in wide area target search are trained to recognize man made targets with minimal target insonification and, during these surveys, log all suspicious targets. In contrast, detail mapping operations require specific vessel speeds and high insonification rates. Only recently has precision equipment and the operational skill been developed for remarkably accurate detail mapping of discrete submerged targets.

Detail mapping is most often used to identify small parts of large targets, but it can also be used to identify small targets. This is particularly valuable in minehunting operations where the mine countermeasures team needs to have as much information as possible concerning the target being scanned. Differing from channel clearance operations using sonar, these operations allow the surface team to pre-classify the target prior to deploying divers or mine neutralizing ROVs.

Detail mapping operations are typically performed using a very experienced sonar team. The ability of the operator/interpreter to recognize predicted and unpredicted components of the image is a key part of the operation.

A highly stable towfish is an important criteria for detail mapping operations. When using a small craft for a support vessel, calm sea states will aid in fish stability. Vessel heave, transmitted directly to the towfish, will degrade a mapping image very quickly. If these operations are attempted on a rough sea, a larger vessel that is less susceptible to heave, helps stability. Reconfiguring the towing assembly to further decouple the towfish from the motion of the vessel may also increase fish stability in rough water.

A carefully determined and closely maintained vessel speed is also important to gain the high resolution needed in the detail mapping operation. Speed must be determined in relation to the target size, water depth, and range used for the mapping process. Vessel speeds that are too high will result in less insonification of the targets constituents than is desirable, and may induce small instabilities in the towfish. Vessel speeds that are too slow can result in towfish path instabilities that will affect the linearity of the target in the final display. Often, several speeds must be tested on the target in order to determine which speed provides the most accurate image of the target's structure.

Water flow over the target site is another consideration during detail mapping procedures. If there are high velocity currents during sonar operations, attendant eddies or small current warbles can induce slight instabilities on the tow cable, towing arm and fish. It is desirable to be close to the target, but stronger current eddies which work against the stability of the fish, may occur in the immediate vicinity of large targets.

In regions where currents are tidal, the best images will be gained during slack water periods. Naturally, these are longest during certain phases of the lunar cycle and surveys planned for those times give the operation a larger window. In areas where the current is not tidal, there may be periods where the currents are minimized. These are the best times to execute detail mapping operations.

Ideally, the detail mapping process involves getting the towfish very close to the target and using the shortest practical range scale. This increases the complexity of any sonar operation not only because the fish could easily collide with the bottom, but also because the site might contain rigging or other projections that could entangle the fish. Precautions taken to avoid collision of the fish with the target will include accurate and timely control of the fish-to-target distances. In relatively shallow water this is usually a simple procedure, but, as the water depth (and associated in-water-cable lengths) increase, it becomes more difficult. This difficulty is primarily due to loss of control of towfish with longer cable lengths and the potentially large cable catenary that is susceptible to changes during the towing process. It can sometimes be helpful to perform trials using the expected cable lengths in a similar environment before the final imaging of the target. Even with its complexities and difficulties, detail mapping is one of the best overall processes for high accuracy imaging of complex underwater targets.

Illustration 140 shows the fishing vessel *Margaret Rose* from above. When the photo was taken, the vessel had stranded on a lee shore. She was salvaged within a few weeks and sailed for several years before sinking in deeper water. In this photograph, the structure of the deck house and the forecastle can be seen.

An example of detail mapping is seen in Record 141 of the same 60 foot eastern-rigged fishing vessel. The acoustic map of the wreck shows the detail of the hull and deck structures. The raised forecastle is intact and so is the house on the stern. Although the mast has fallen since she sank, the record indicates that there is some rigging on the deck of the vessel. In the acoustic shadow of the wreck, two dark spots show the presence of scuppers or other type of breach in the bulwark. The splash guard on the bow is also intact. Where many short range detail mapping operations will use the higher 500 kHz frequency, this mapping operation had the best results using 100 kHz. This is due to the soft wood construction of the target vessel's hull. In this record, the overall gain of the sonar system was set relatively low so the interpreter could discern the finer details of the strong target reflectors.

Record 141: The record above is a detail map of the wreck of the *Margaret Rose* after it sank in deep water. The sonar image shows the raised forecastle, deck area and the eastern-rig construction.

Illustration 142 is a photograph of the single stack freighter *Port Hunter*, used for transoceanic commerce. She was built in 1906 and had a tonnage of 4,062. During World War I, while carrying supplies for the front lines in Europe, she was in a collision with a tugboat off the coast of New England. The ship was run onto a shoal to prevent her form sinking in deep water. Over the decades some minor salvage attempts were made on the vessel, but she has remained largely intact. Record 143 shows a detail map of this 100 meter long wreck one-half-century after its sinking. Time has taken its toll on the ship's structure, but sonar mapping points out important features about the wreck.

Illustration 142: The single stack freighter *Port Hunter* sailed for Europe with supplies for the American troops in 1918. Before she left the continental shelf of New England, however, she collided with a tug boat and sank, fully loaded. The wreck has survived several salvage attempts during the decades since she sank.

Record 143: This detail map of the *Port Hunter* shows the position of several hold openings and a major structural failure in the midship region of the wreck.

The hold openings and the hull are, for the most part, intact with the exception of a major structural failure amidships. The split of the hull resulted from the vessel being sunk on a shoal which eroded away, leaving the stern unsupported. There are no bridge or deck houses evident and some areas on the deck are debris covered.

Objects on the sea floor sometimes become buried in a short time, particularly in areas where sediment transport is high. There is very little left that shows above the sand where the four-masted, schooner-rigged *West Virginia* sank, as seen in Record 144. Once she settled to the bottom, the sea took its toll as the upper decks fell in or were swept away, and finally the sides were worn down to just above the sediment. What shows here is the frame tops and planking of the vessel running from bow to stern and her keelson running through the bilge. A cross section of the ship, and the portions visible in the sonar record, is shown in Illustration 145.

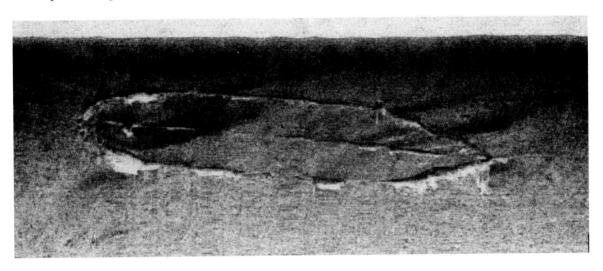

Record 144: The 75 meter-long *West Virginia* sank in the late 1800's at an exposed coastal location. The wreck slowly disintegrated until all that remained was the lower portion of the ship buried in the sand. This sonar record shows what remains of the vessel above the seafloor. The ship's keelson protrudes from the sediment throughout her entire length. (see Illustration 145)

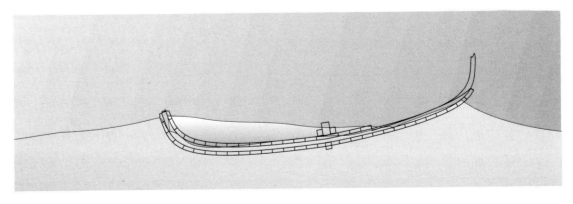

Illustration 145: The remains of the *West Virginia* represent classic wooden shipwreck sites. After one-century at a location exposed to open ocean energy, a wooden wreck will have lost her rigging, fittings, upper decks, and frame tops to the environment. This diagram shows what the site of the *West Virginia* looks like in cross section.

A small sand wave can be seen crossing the wreck. Like the *West Virginia*, ships sunk in active geological areas alternately become buried and uncovered over time. The detail mapping process outlines areas where a ship is buried by defining slight changes in the seabed as the buried portions of the wreck alter sedimentation and water flow around the site.

Careful mapping can provide significant information about the structure and design of sunken targets. Illustration 146 shows the outboard elevation of an L-class submarine designed and built in the early 1900's. The drawing shows the position of the cutwater at the bow (A), the torpedo hatch, a personnel hatch, and the gun mount on the forward section (B). Amidships is the conning tower (C). The conning tower fairwater is shown with dashed lines. On the stern deck, a personnel hatchway is shown (D). The stern of the L-class submarines was raised (E) from the keel starting about half way aft of the conning tower.

One of these submarines, the *L-8*, was used for the testing of magnetically influenced torpedoes during early ASW efforts by the Navy. Although magnetic and pressure sensors are standard triggering mechanisms today, the early systems had prototype problems. Illustration 147 shows the firing of the new torpedo under the submarine during a test. It passed under the target without exploding. A second torpedo functioned properly and put the submarine on the bottom.

Record 148 shows the sub as she has lain for over 50 years on the bottom. Examination of the record shows buckled hull plates and an overall deterioration of the upper sides of the hull. Notable in the sonar record are the cutwater (A), torpedo hatch (B), gun mount and personnel hatch area (C), and conning tower (D). The remains of the conning tower fairwater are only slightly visible (E). In the stern, a personnel hatch (F), the rudder and rudder supports (G) are most notable in the target shadow. The upward rake of the submarines stern is indicated by the separation of the shadow from the target.

Large target sites can also be mapped to provide accurate dimensions and conditions of the site. Illustration 149 shows the 6,888 ton screw cruiser *Yankee*. She was built in 1892 in Newport News, Virginia, and was armed with ten 5 inch

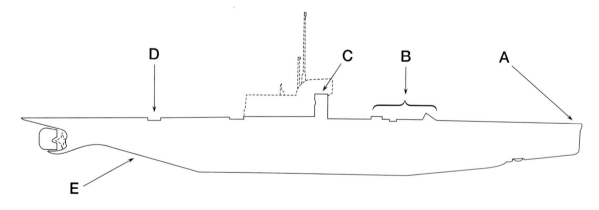

Illustration 146: Early submarines had hull designs very different from those in use today. The outboard elevation of the *L-8*, a subject of detail sonar imaging in Record 148, is shown above. The drawing shows the position of the cutwater (A), hatches and gun mount (B), and conning tower (C). The conning tower fairwater is shown with dotted lines. On the stern deck, a personnel hatchway is shown (D). The stern of the L-class submarines was raised (E) from the keel.

Illustration 147: The *L-8* was used for testing a magnetically fired torpedo prior to World War II. During the first test the torpedo failed to detonate and passed harmlessly under the doomed sub's hull. A second firing was successful in sinking the submarine. *From the collection of Professor Henry C. Keatts.*

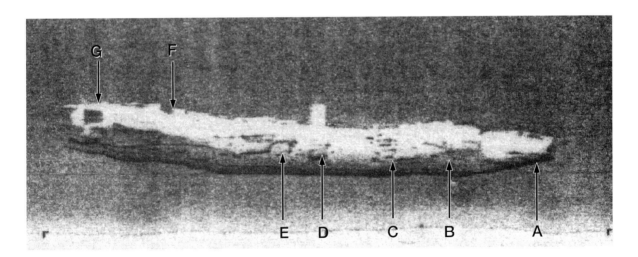

Record 148: The *L-8* as it lies on the bottom after more than half a century. The hull is badly deteriorated but still shows some interesting features such as the cutwater (A), torpedo hatch (B), gun mount and personnel hatch area (C), conning tower (D), and the remains of the fairwater (E). A personnel hatch (F), the rudder and rudder supports (G) are most evident in the shadow.

Record 149: The naval cruiser *Yankee* was built in 1892 and served in the Spanish-American War. She was sunk in 1908 near shipping lanes and proved unsalvageable. The government dynamited the wreckage to eliminate it as a hazard to navigation. *From the collection of William P. Quinn.*

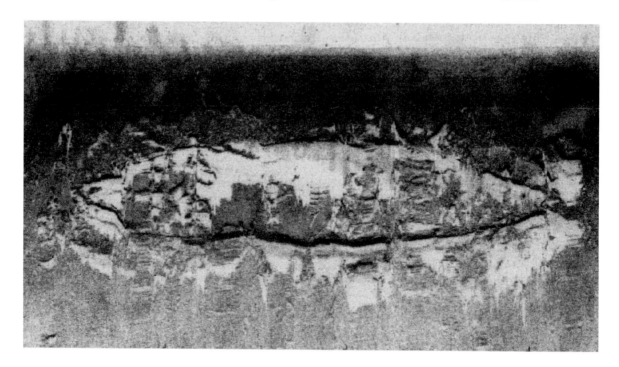

Record 150: This detail map of the remains of the *Yankee* shows the extent of the government dynamiting efforts. The ship was blown down to within 5 meters of the bottom, strewing portions of the heavy steel plating over a wide area. The sonar record shows the extent of the remains of the hull and the highest parts of the wreck.

rapid fire guns, six 6 pounders and two Colt automatics. She participated in numerous naval battles as a victor during the Spanish-American War, and in the early part of the 20th century, served as a training ship. In September 1908 she went aground off the coast of New England. Salvage attempts were partially successful in floating her, but she was abandoned after several months. Stripped of her armament, she was allowed to settled to the bottom. The Government expended tons of explosives to bring the least depth of the wreckage close to the bottom and prevent her from being a hazard to navigation. When the blasting was complete, the remains of the *Yankee* rose no more than 20 feet off the bottom.

The detailed side scan Record 150, of the remains of the Naval cruiser, shows the extent to which the government clearance efforts mangled the wreckage. Most of the decking was blown in and the ship's sides were cut down with the buckled hull plates falling both inboard and outboard of the vertical hull.

A detail map of a wreck in the condition of the *Yankee* cannot provide information on previous components of the ship such as bow from stern, but it can tell the surveyor which portions of the wreck rise higher than others and where the more complex portions of wreckage might be found. It will also outline the areas containing parts of the ship that were blown away from the hull and lay disconnected from the main wreckage.

On December 7 1941 one of the most historic wartime actions took place when the Japanese attacked Pearl Harbor on Oahu. In an air attack on the island and the Naval vessels congregated there, the Japanese sank the main battle force of the Pacific Fleet. The vessels sunk included: *USS Oklahoma, USS California, USS West Virginia, USS Nevada, USS Arizona* and the retired battleship *USS Utah.* Fortunately the sinkings took place in shallow water which allowed the possibility of salvage.

The *USS Nevada* was refloated and repaired in dry dock. The *USS California* was pumped and refloated in March 1942. The battleship *USS West Virginia* was raised and entered dry dock in June 1942. The *USS Oklahoma,* which had capsized at its berth, was righted with the use of twenty-one electric winches mounted on shore and 40 foot long struts welded to the overturned hull. She was then removed from the harbor.

The *Arizona* was so totally devastated that the hull was never salvaged. Eleven hundred and seventy seven men died in the sinking of this ship alone. In the months following the sinking her guns were removed and she was stripped of all usable equipment down to the main deck.

The *USS Utah* was hit early during the attack by two aerial torpedoes. She sank within minutes. As the huge ship filled, she turned 180 degrees from upright. Timbers on deck shifted, killing some men and trapping others below.

The *Utah* was deemed salvageable and, like the *Oklahoma,* electric winches, cables and struts were applied to the hull in an effort to right her so she could be towed to dry dock. Salvage efforts culminated early in 1944 but as the winches groaned and the ship began to roll upright she suddenly started to slide along the bottom towards shore. The vessel had come around to about 40 degrees from upright by that time but the winches were halted while the salvage master reconsidered options. It was finally determined that continued salvage would be too costly and in March of the same year the work stopped leaving the ship on her port side at a 38 degree list.

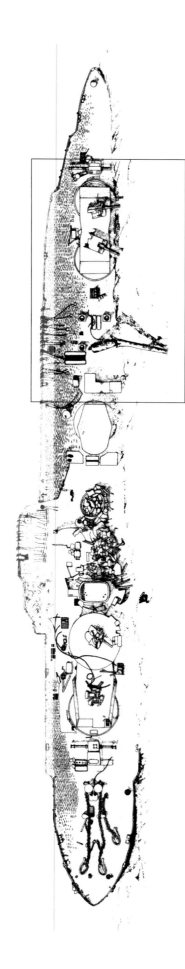

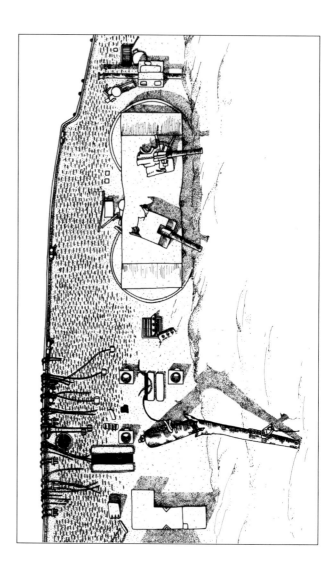

Illustration150 A: When the Japanese attacked Pearl Harbor, the *USS Utah* was hit and sank within minutes. She is currently a war memorial under the jurisdiction of the U.S. Navy and the National Park Service. Precise underwater surveys of the wreck were made under the direction of Dan Lenihan and Larry Murphy of the NPS, and Commander Dave McCampbell of the US Navy. This intensive mapping operation resulted in the drawing above. *Courtesy National Park Service*

Record 150 B: The record above is a detailed sonar map of the wreck. Prominent are: the anchor chains (A), the forward gun turrets (B), the remains of the elevated superstructure (C), a 1.1 inch quad antiaircraft gun (D), the remains of the ships stack (E), a raised turret mount (H) and deck components (F, G, and I). Further aft, the main mast (J) rests on the sediment, and the two sternmost turrets still have 5 inch guns mounted (L). The housing of the aft gun is missing. Other discrete items include hatchways, winch and stern bits (K, M and N).

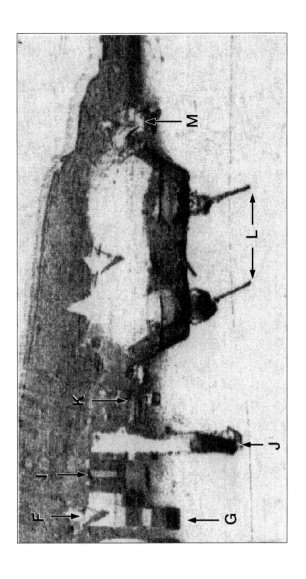

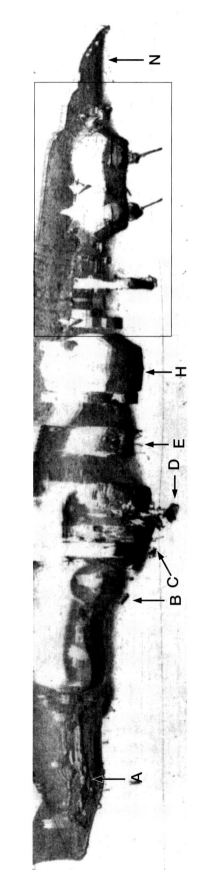

Detail Mapping

Both the *USS Utah* and the *USS Arizona* are currently war memorials in Pearl Harbor under the jurisdiction of the National Park Service and the US Navy. In a joint operation of the NPS and the Navy, detailed underwater surveys were made of both wrecks during the 1980's. Under the direction of Dan Lenihan and Larry Murphy of the NPS, and Commander Dave McCampbell of the US Navy, divers made detailed drawings of the ships. Illustration 150A shows the results of this intensive mapping operation which included measuring and recording both large and small components of the Utah. Record 150B is a 500 kHz side scan sonar detail map of the wreck. The record displays some of the finer details mapped by the divers. Even though side scan may not easily provide images of very small objects that can be mapped by hand or with a positioning systems like Sharps™ (Marine Telepresence), it can provide excellent detail of complex target components relative to one another.

From a mapping standpoint, the *Utah* is interesting because she was not completely stripped of her superstructure after sinking. She is one of the last "Super-Dreadnoughts" in existence today. When she was built, the *Utah* sported ten 12-inch, and sixteen 5-inch, guns. Her armor plating is 12 inches thick and she carried two submerged 21-inch torpedo tubes. The *Utah* is still laying on her side partially buried in the harbor floor.

Comparing the diver's map and the sonar record which was made several years later, it can be seen that the port side of the wreck is becoming buried deeper with time. Several components on the wreck are notable in the sonar record. The anchor chains (A) lay on the deck at the bow. The forward gun turrets (B) cast a long shadow on the starboard deck. Aft of these are the remains of the elevated superstructure, damaged both in the attack and during later salvage attempts (C). Still attached to the hull but only partially exposed from the mud is a 1.1 inch quad antiaircraft gun (D). Behind the superstructure rubble are the remains of the ships stack (E). The large structure just aft of the stack is a raised turret mount (H). Near this stern turret, numerous discrete deck components can be distinguished as mapped out by the dive team (F, G, and I). The main mast is broken at the base and rests on the mud (J). One striking feature of the sonar record is the image of the two sternmost turrets which still have 5 inch guns mounted (L). The difference in the angle of the barrels between the record and the drawing is likely to be due to the aspect at which they were acoustically imaged ('first-return, first-displayed' phenomena). It can be seen by examining the sonar image of the two five inch guns that the housing at the base of the forward gun is intact, while the that of the aft is missing. Other discrete components visible here include hatchways, the aft winch and stern bits (K, M and N).

The detail mapping process can be applied to a number of underwater targets other than mines and shipwrecks. Almost any target of interest can be mapped in a permissible environment. For discrete objects such as dam faces or dredge areas, sonar can be used as a large scale, high resolution imaging tool. After these sonar images are made, the surveyor will often use an ROV or diver for very close inspection of specific areas defined by the sonar. These subsequent tools provide very high resolution images of trouble spots, but because of its wide and variable ranges, the side scan is a useful macro-structure tool for assessing the entire target as a whole. A sonar survey shows on a large scale, damage to concrete or steel used in the makeup of bridges. Record 151 shows two types of such abutments: a single-support footing and a double footing.

Recent concern over the danger of aging roadway bridges in the United States has brought in the use of side scan as a primary tool of inspection on bridge abutments and footings in rivers, harbors and other waterways. Record 152 shows a concrete bridge abutment which was placed on a platform footing for support. The bridge carries an average of 1200 cars per day and is over 60 years of age. The above water sections are deteriorated and the submerged components have been highly suspect.

Damage to the concrete footing of the support can be seen in this record. In the damaged area, reinforcing materials are exposed to the underwater environment increasing the rate of deterioration of the structure. After sonar examination of this bridge support, divers were deployed to examine damaged areas more closely.

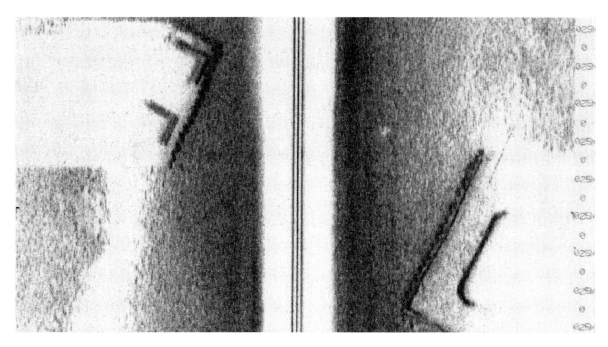

Record 151: Bridge abutments are constructed in a variety of forms. They can be accurately imaged by sonar in preliminary surveys before diving operations. Both the single and the double abutments in the record above are supported on reinforced concrete pads.

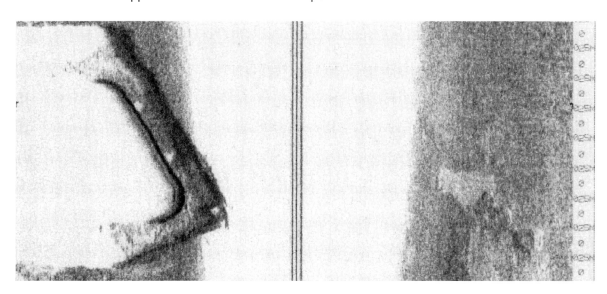

Record 152: The detail sonar image above shows severe deterioration of the concrete pad upon which the bridge supports rests. The concrete has spalled away from this footing resulting in increased breakdown of the structure. This particular bridge is heavily deteriorated, yet still carries over 1200 cars daily.

INVERTED IMAGES

Unusual phenomenon not typically encountered during larger scale survey operations can occur during detail mapping operations. At close range, the detail mapping process will occasionally produce a record that can be misleading when trying to determine the orientation of the target on the seabed. The record can even produce an inverted or "mirror" image of the target. Although the details of the target are not affected, these reversed images are often confusing. Recognizing that sonar imagery is based on the time of arrival of reflected sound, it is clear that the first arriving pulses will be the first displayed. Sometimes, if a vessel is lying on its side and the towfish is towed very close to the target, it may be imaged in the reverse. The "mirror image" phenomenon in sonar most often occurs when scanning a target with considerable relief towards the towfish. It will occur when the ratio of towfish-target distance to target-relief approaches 1:1 and the vertical bearing to the target is small (<45°). Illustration 153 shows this geometry.

Other less dramatic illusions can occur when mapping targets. In the case of the *Margaret Rose* (Record 141) the image gives the illusion of the vessel lying on its starboard side when, in reality, it is perfectly level.

Another example of this inverted imagery is seen in some of the constituents of a target as well. With the towfish very close to part of a target that protrudes up into the water column, the first-return, first-displayed phenomena make unusual images. An examination of Record 143 of the shipwreck *Port Hunter* will show what appears to be the starboard hull at the bow imaged clearly from the deck to the sand below. However, examining the sonar geometry, it can be seen that the sonar would not have been able to image the starboard side of the hull, since the towfish was flown on the port side of the wreck. Although this image is of the wreck's port side, it is recorded and displayed on the starboard side of the image.

As is the case with other sonar phenomena, careful study of the imaging geometry (towfish-target-surface-seafloor relationship) will often provide the true orientation of targets that appear to be unexpectedly configured.

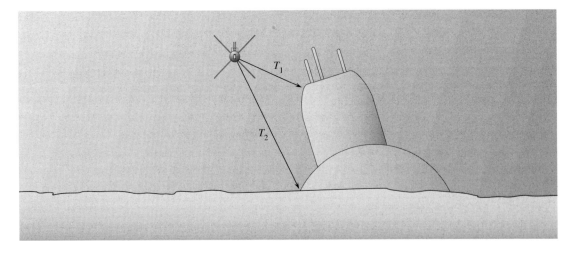

Illustration 153: The first-return, first-displayed phenomenon can create unusual sonar images. Scanning targets that have high relief in relation to towfish-target range may display objects in improper positions in the final two-dimensional record. In the above sketch, returns from objects along T_2 will be printed at a greater range than those from T_1. If a detailed image appears to be confusing, examination of the sonar geometry will often provide clues to the reasons for the imagery.

Chapter 11:
Sonar Targets

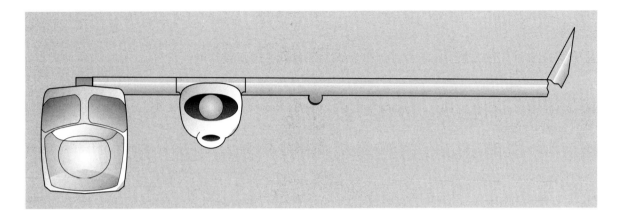

SHIPWRECKS

Historical and modern shipwrecks are a common target for side scan sonar search operations. The side scan is often a first order tool for the surveyor searching for a shipwreck when he needs to identify it remotely. After the target is located by wide area search operations, side scan sonar helps by providing accurate images of the makeup of the wreck. Orienting a wreck, such as determining bow from stern, outlining the ship's bridge structures, hold openings, or areas of structural failure, is often performed by detail mapping (Chapter 10). But even without employing the specialized procedures required for detail mapping, the sonar provides considerable information about the target with each pass and each image.

Search operations for different shipwrecks are configured differently and depend upon the size and makeup of the target wreck. These operations must be carefully planned in order to cover the search area without wasting time or missing the target.

One example of a sonar record illustrating different types of shipwrecks is Record 154. This record was generated during a search for 18th century sailing vessels known to have been sunk during the Revolutionary War. The record shows three major targets, two of which are shipwrecks.

The target containing two parallel lines (**A**) on the port channel, is the wreck of a modern barge. The sonar operator readily sees that the size of this target is consistent with man-made objects and it is a strong reflector containing straight parallel lines. Identification of this type of target requires shorter range inspection with sonar.

Less obvious to the sonar operator may be the large, shadow producing target (**B**) on the starboard channel. Although this target lacks straight lines, the size, shape, height (evidenced by the shadow), and its inconsistency on this bottom, would indicate to the prudent sonar operator that the target should be examined more

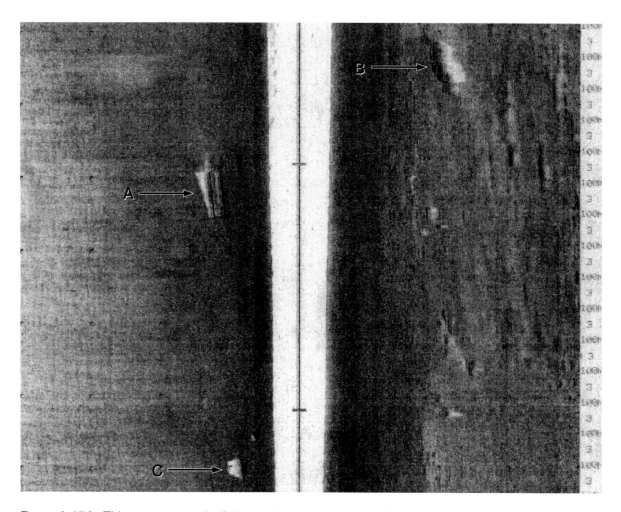

Record 154: This sonar record of the seabed shows several targets. A shipwreck (A) is readily recognizable due to its straight lines and distinct shadow. A target (B) is clearly an anomaly but appears more amorphous than (A). It is an older shipwreck's ballast pile. A shipwreck search operation would examine B more closely due to its length and height off the seabed. Target (C) generally appears too high and short to be a shipwreck and would take low priority as a potential target of interest.

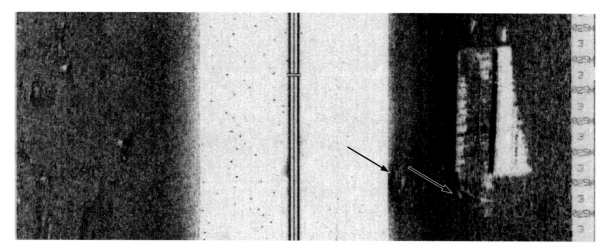

Record 155: Here, target (A) in Record 154 is examined more closely. Although the barge appears to have lost some of its cargo (arrow), in reality this wreck sank on top of the remains of a ship sunk in 1779. The older vessel's ballast pile can be seen near the first bottom return and under the end of the barge.

closely. Closer inspection with a sonar and finally an ROV (Remotely Operated Vehicle), confirmed that the target is the floor timbers and ballast pile of a 19th century sailing vessel.

A third target (**C**) visible in this record provides a strong return and has a somewhat amorphous shadow shape without straight lines. It is not large, and although the target is a rock, only closer inspection of the target would confirm this. The length and width of the target appears to be approximately equal to its height. This is one clue indicating a natural rather than man made target.

Even in the long range record of the barge, (**A**), it is seen from the shadow that one end is higher off the bottom than the other. Record 155 confirms this and shows more detail of the actual construction of the craft. The sonar record shows that one end of the barge is damaged (the portion of the target raised high off the seabed) and offers the indication that some cargo has spilled. After inspection, it was determined that this wreck sank almost directly on the remains of a Revolutionary warship, and the "cargo" of the barge is actually the ballast pile from the other wreck. Portions of this ballast pile are also seen directly under the towfish on the starboard channel.

During target location efforts, particularly those involving shipwrecks, it is very helpful to know the dimensions of the particular target in order to determine the optimal tow speeds and range settings because insonification rates are important for detection and recognition. Record 156 was generated during a search for a sunken tug boat. The bottom conditions were smooth and flat, providing a good search area. This allowed the use of higher tow speeds and longer ranges than could be used on a more cluttered seabed. The record shows that the trackline of the towfish happened to be 90° to the long axis of the target. This aspect presents the smallest profile for insonification and makes detection and recognition more difficult. However, since speeds and ranges were carefully selected in the planning stages of the survey, the target is still recognizable as a distinct seabed anomaly warranting shorter range inspection.

Record 157 shows the same target on close range. It has the size and features of the target of interest. Visual inspection proved that it was the missing tugboat. Note in the short range record however, the rounded stern common to tugboats of this design and the shadow which falls away from this portion of the target. The stern of the tugboat is raised from the seabed and this allows some insonification of the area just below the stern. These features in the record helped the sonar team prioritize this target.

The surveyor targeting shipwrecks with side scan sonar can benefit by imaging the wrecks from different angles and ranges. Since many ships, particularly modern ones, will rest on one or another side of the keel, they frequently have a slight list after sinking. This list can be accentuated by currents or other environmental factors over time.

Record 158 shows the remains of a modern fishing dragger lying on the bottom. In the sonar record, the vessel appears to be lying upright. The bow of the vessel, including the bridge, is to the left. The stern shows the gallus frames and net reel which are most notable in the shadow. At the midship's area, sand is beginning

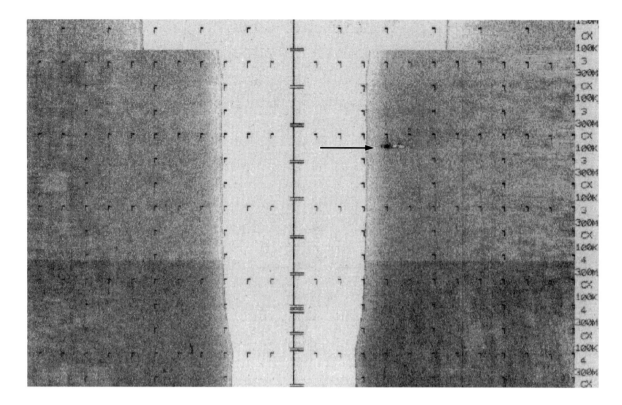

Record 156: This record was generated during a search on flat seabed. The object of the search was a small tugboat. High tow speeds and long ranges were allowed due to the smooth, even nature of the seafloor. The first contact with the target was end-on (arrow, see Record 157).

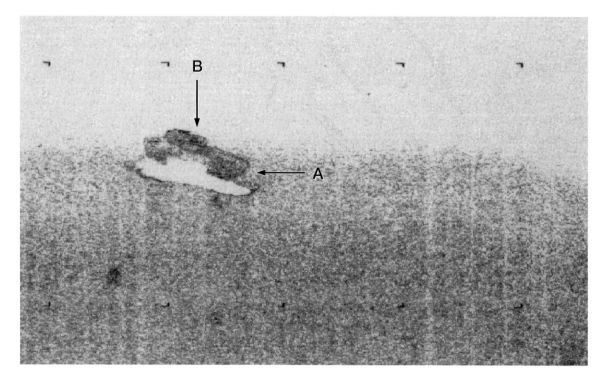

Record 157: The target scanned in the previous record is shown here on a shorter range. The image is consistent with the object of the search showing a high rounded stern (A) and forward deck structures (B).

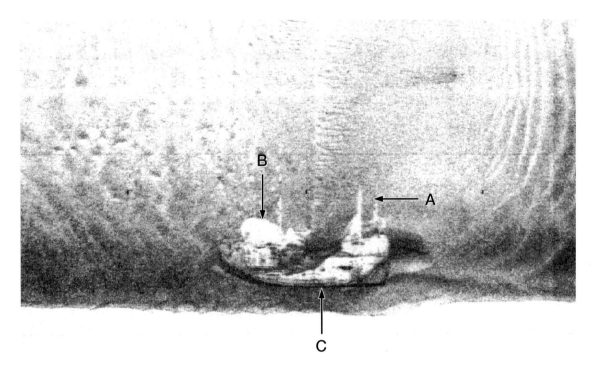

Record 158: Sub sea targets can be imaged using a variety of sonar geometries. This record of a steel fishing vessel shows a top view . The shadows indicate the presence of stern equipment (A) and forward house (B). A large sand wave is crossing the wreck covering much of the midship area (C) (see Record 159).

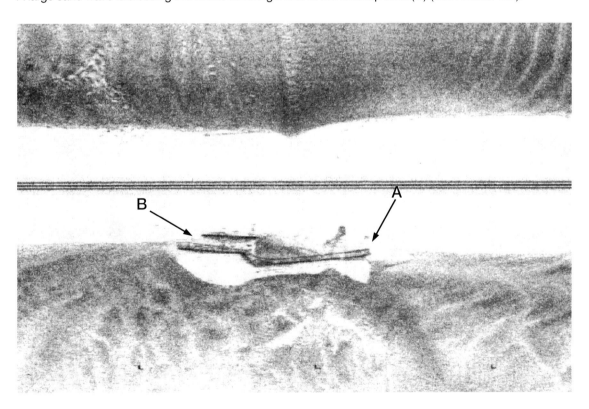

Record 159: The target in this record is the same as that in Record 158. Using different sonar geometry the target is shown with an apparent side view. Although the stern equipment is not as visible in this record, the railing around the gunwale (A) and the forward house (B) are distinctly imaged.

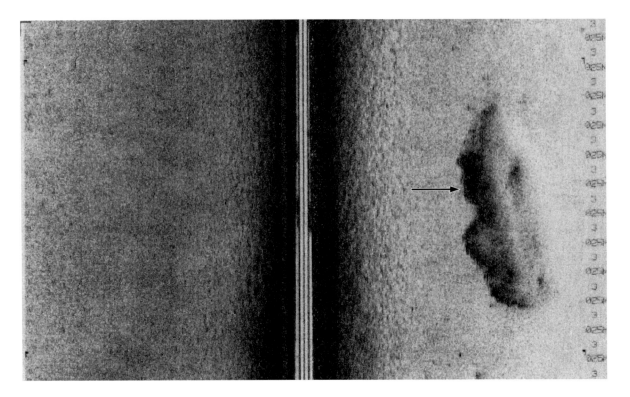

Record 160: Older shipwrecks are more challenging to differentiate from natural anomalies. This target (arrow) is all that remains of a wooden vessel carrying lime, which was sunk in 1898. Although the soft seafloor sediment shows no evidence of a shipwreck upon visual inspection, sonar outlines the wreck because of the change in backscatter level of the sediments near the wreck.

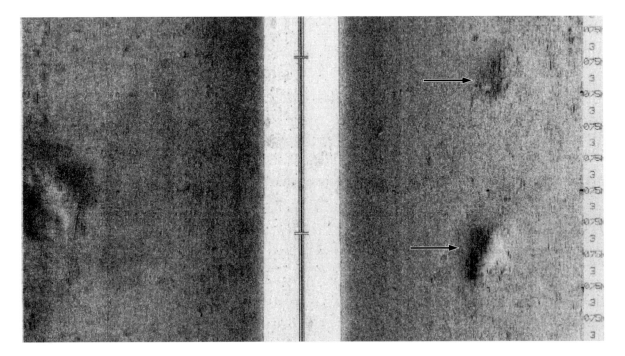

Record 161: During a search for two-century-old shipwrecks the two targets (arrow) in the record above were pinpointed and visually inspected. They are ballast piles from ships sunk in the 1700's. Although they have little relief from the seabed, the targets backscattering properties allowed rapid operator recognition.

Illustration 162: The *Herman Winter* struck the dreaded rocks at "Devils Bridge" off the coast of New England in 1944. Salvage attempts lasted for weeks until it was determined that she would be a total loss and all efforts were abandoned.

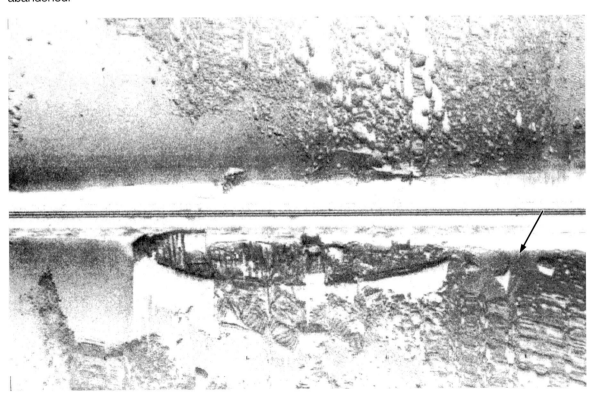

Record 163: The wreck of the *Herman Winter* was aerially bombed for practice during World War II. Demolition was particularly extensive in the stern of the vessel where only the outer hull plates remain (arrow).

to flow over the wreck and fill it. This sand ridge is noticeable in the outer ranges of the record. In reality, this wreck is listing to starboard. This is seen in Record 161 which is another image of the wreck made during the same survey .

Record 159 shows the details of the hull more clearly for identification purposes. The railing around the entire vessel is visible along with the forward "house" or bridge. The sand ridge approaching from the starboard side of the vessel can be seen. The net reel is not as apparent in this image as it is in Record 158. These records demonstrate the value of imaging targets from a variety of angles.

While the sonar operator depends upon modern wrecks to provide a predictable profile on the seabed, older ones are more difficult to detect and recognize. Older wrecks are often substantially deteriorated, and after time, present only low profiles. Side scan sonar is designed to detect any objects protruding off the seafloor, but it sometimes can detect changes in the seafloor itself caused by a shipwreck. Record 160 shows the site where a lime schooner sank in 1898. The wreck occurred in a shallow harbor and the area has since filled in with mud and silt. A diver inspection of the area showed no change in depth at the site. Probing the silt over the wreck, the divers found a depth of refusal at 10-12 centimeters. This harder layer appears to be the cargo of lime carried by the schooner. The amorphous target in this record shows the actual cargo. It is detected by the differing makeup of the seabed caused by the buried lime.

Very old shipwrecks will often leave some telltale trace of their existence even centuries after they sink. In Record 161, ballast piles from shipwrecks sunk in the 1770's can be seen on the starboard channel. Although the ballast piles have very little relief from the seabed, and have not caused a regional change in the sediment, their reflectivity, or backscattering properties, are very different from the surrounding topography, making the targets recognizable in the sonar image.

Modern wrecks in rugged bottom topography are as challenging to locate as older ones; but once they are recognized, it is a straightforward process to image them. Sonar is often used as a tool for determining the shallowest area in the region. Other details about the condition of the wreck can also be established.

The *Herman Winter* was built in 1887 and plied the coastal waterways of the eastern United States for half a century. After striking a rock ledge in New England in the spring of 1944, she was deemed a total loss. The ship is shown undergoing unsuccessful salvage attempts in Illustration 162. World War II was raging, and the airborne military found the *Herman Winter* to be an attractive target for practice bombing. The result was demolition of much of the wreck's structure.

The side scan data from the site, shown in Record 163, reveals the extent of the damage experienced from the aerial bombing. Of particular note is the damage to the stern of the vessel (arrow). Only a small amount of material is visible in the stern as seen at the curved hull plating.

AIRCRAFT TARGETS

Because side scan sonar is the best method of imaging large areas of the seabed, downed aircraft are a common target for sonar operations. All the major aircraft recoveries in the 1980's, including the Korean Airlines Flight 007, shot down over the Sakaline Islands; the NASA Challenger Disaster in January of 1986; and the South African Airways 747 in 15,000 feet of water near Mauritius, employed side scan sonar to locate the aircraft components. Hundreds of lesser known projects also used the wide swath capabilities of side scan to hunt for aircraft lost underwater.

The type of sonar search for lost aircraft ranges from a brief one-day operation to pinpoint the exact location of an aircraft that impacted at a known location, to a large-area, long-term search that might last weeks or more. The duration of the operation depends upon two factors. The first relates directly to the positioning accuracy of the point of impact and the second relates to the condition of the target on the seafloor.

Often, large aircraft are tracked by Air Traffic Control (ATC) radar when they are in the region of a radar center. Within the coastal zones of industrialized countries, this is a continuous monitoring system that tracks aircraft location at altitudes above 800-1200 feet. When the larger aircraft go down over coastal waters, side scan will usually pinpoint the location of debris quickly. Used properly, side scan is also able to map out any debris fields that may exist near large portions of wreckage.

Crashes of large aircraft that fly out of range of ATC radar, small planes, and rotary wing aircraft that may not be tracked by radar, are not often precisely positioned. These accidents sometimes require long search operations that eliminate large areas before determining the location of the aircraft.

Data that supports the planning of this type of search includes the position of floating debris found after the accident. During the planning phase it is very important to gather all the data on any debris that relates to the incident. The amount of scientific data on the oceanographic processes that effect the drift of floating debris, including wind driven currents, tidal currents and storms has become significant in the past few decades.

In the 1980's and 1990's there have been many aircraft search operations using the record of debris location to help define the search area. In one instance, an aircraft was lost and the last noted position on the ATC radar was the only clue to its location. That point was established as the center for the search. The search area was widened when the first ten square miles did not hold the target. Twenty-one days after the incident, one small piece of debris washed up on a beach over fifty miles from the search area. Computerized drift analysis on that single bit of debris suggested the search area be shifted to another location thirty-five miles away. The target aircraft was located in the new location. This is a good example of the value of floating debris and its position. The best data is from debris that is still floating when it is found, because it can be backtracked from its present position against time more accurately. If debris is found washed ashore, it is important to try to establish the time that it washed up on shore to assist the analysis effort.

The condition of aircraft on the seafloor depends upon several factors including speed at impact, size of the aircraft, and bottom activity in the region. An aircraft that impacted the surface of the water at cruising speed is likely to be more broken up than one whose pilot made some attempt at a belly-down landing on top of the water. Higher speed aircraft such as those that travel at speeds in excess of 175 miles per hour often have a high stall speed and are likely to be more broken up. Larger aircraft often fall into the category of high stall speed, but they are also constructed so they are more susceptible to breakup on water impact than are smaller aircraft. The small fixed and rotary wing aircraft are often found in better condition on the bottom.

The activity of currents, storms and fishing in the region will also affect the condition of the craft. When aircraft crashes occur in regions heavily fished by bottom trawlers, it is important to begin search operations immediately. Fishing dragger equipment is very destructive to a sunken aircraft, and an intact craft will be easier to recognize than a debris field. There have been occasions when dragger nets caught on an aircraft have pulled a large part away from the main portion of wreckage and towed it for miles. In some cases the vessel captain never knew he had caught onto, and moved, part of a plane.

As in other sonar operations, planning is very important but for aircraft searches 'time is of the essence.' Radar information should be gained, debris locations should be researched, and drift studies performed before the data becomes too old. Search operations should resume immediately if there is a need to recover any part of the aircraft or its contents.

During some types of search operations, the operator rarely knows what shape or condition the target will be in prior to contact. During aircraft searches, the target's sonar image may not resemble an aircraft, and targets of all shapes and configurations must be examined. Aircraft lost underwater take varied forms on a sonar record, ranging from an intact image of the craft to a widespread debris field of small pieces. Even if an aircraft is expected to have impacted the water at a low speed, deterioration from natural or other causes can easily break it up and spread it out. Craft that have sunk in areas of high sediment transport can be buried in a matter of weeks.

The aircraft that do not disintegrate upon impact often provide coherent images for the operator. Sonar Record 164 is of a Douglas AD-5 Sky Raider that crashed and lost its engine on impact. The rest of the craft remained intact. It is readily identifiable from the sonar record, but it has become largely buried by sand. In searches for aircraft where the target is expected to be this coherent, ranges and tow speeds are determined by carefully evaluating the size and shape of the target.

Record 165 shows the remains of a recently sunken aircraft that has disintegrated either during, or sometime after, impact with the sea surface. There is no portion of this target that indicates that the debris field might have come from an aircraft. The target consists primarily of aluminum sheet metal and aluminum framing, but does not demonstrate the sharp edges and shadows commonly associated with this type of structure. This is partly due to the high altitude of the towfish when the target was imaged.

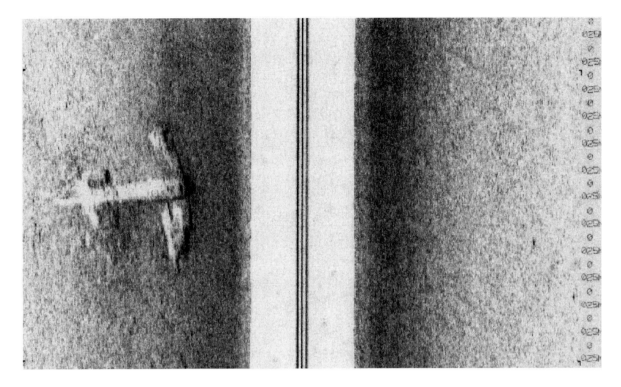

Record 164: The aircraft in this record is a Douglas AD-5 Sky Raider. It is remarkably intact even though the engine was lost on impact. The sediment is beginning to cover the aircraft's wings, but the aircraft is still readily recognizable.

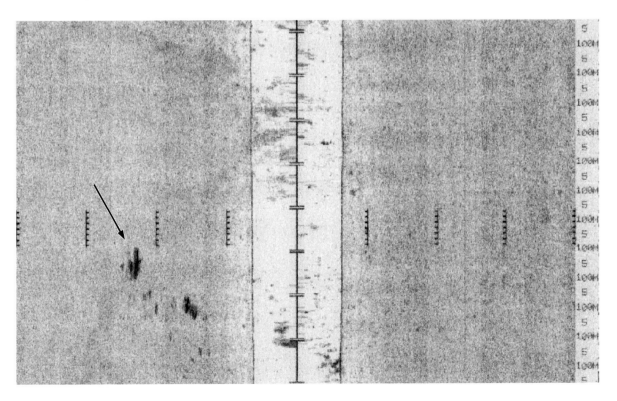

Record 165: This record shows a debris field consisting of components of a multi-passenger, fixed-wing aircraft. It disintegrated during or sometime after impact with the sea surface. Although most of the aircraft is contained in the debris field (arrow), it is not easily recognized as an aircraft such as those in Records 164 and 169.

Record 166 shows the remains of a recently lost aircraft. The plane was of a twin-engine, fixed high-wing design. Upon impact, the aircraft broke into several pieces. This record shows one piece which consists of the cockpit, one wing and one engine as detailed in Illustration 167. The outer wing tip has been bent upwards and this structure is identified in the shadow. When this target was first located on long (300 meters per side) range, it presented a small, but recognizable image. The target was then surveyed at shorter ranges to gain this image of the wing and cockpit. The other wing of this craft was located one and one-half miles away. It is shown in Record 168. The large dark mass in the water column casting a thin shadow on the seafloor is a school of fish in the water column just under the towfish. An interesting feature of this record is that although the wing's skin was intact, the sonar penetrated this outer layer and was reflected by the frames inside, giving a "ribbed" effect to the image. This sound penetration will occur at times, depending on the sonar angle of incidence and the makeup of the target material.

Record 169 is of a Grumman AF-2 Guardian, four passenger ASW aircraft. The structure of the plane has broken down with one wing completely buried in the sediments and the tail section broken away.

Without the knowledge of the condition of a target, it is sometimes difficult to choose sonar ranges and vessel tow speeds that will generate a recognizable image of the target. During searches of this kind it is better to be more conservative in choosing these search parameters.

For smaller aircraft, tow speeds should be maintained at 3-4 knots or less when using longer range settings to allocate enough sonar pulses on the target for recognition of small debris. When running at shorter ranges these speeds can be increased proportionally.

Most debris fields from private aircraft will be detected by ranges out to 150 meters on a flat seabed. However, when the operation is performed in rugged bottom terrain, sonar ranges must be reduced to maintain recognition of the man-made debris.

Different aircraft will present different images on sonar records depending upon their original size, and their post-crash condition. Record 170 was generated using a range of 200 meters per side and a tow speed of 2.5 knots. The target is a small fixed wing aircraft (Cessna 152), largely intact with a small debris field around it. The image of the target, generated by the sonar, is too small to be readily recognized by an operator especially given the possibility of other bottom and water borne targets that may exist. Record 171 is of the same aircraft, generated at the same tow speeds, but using a 150 meters per side range scale. Here, the target is more readily recognizable. The target is closer to the tow path in this record, but the major increase in target clarity is due to the change in range scale.

In planning aircraft searches the sonar team should examine all of the likely scenarios for the accident and determine the worst case for the condition of the subsea target. This will provide the team with the correct search parameters for rapid recognition of the target when insonified.

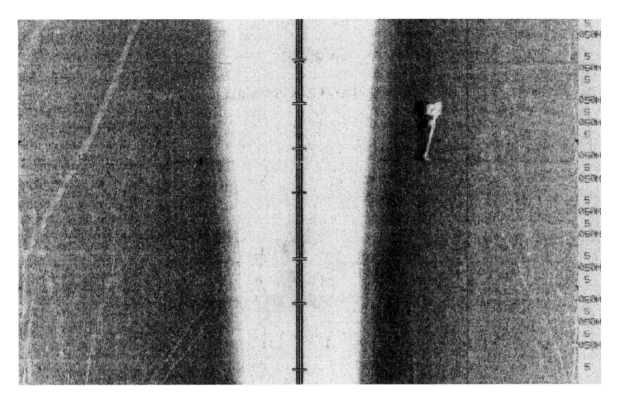

Record 166: This record shows the cockpit, one engine and one wing of a recently lost aircraft. The components of the plane that are in this record are shown in Illustration 167.

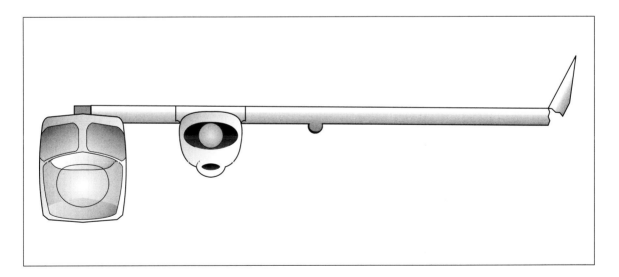

Illustration 167: This outline drawing shows the components of the aircraft in Record 166.

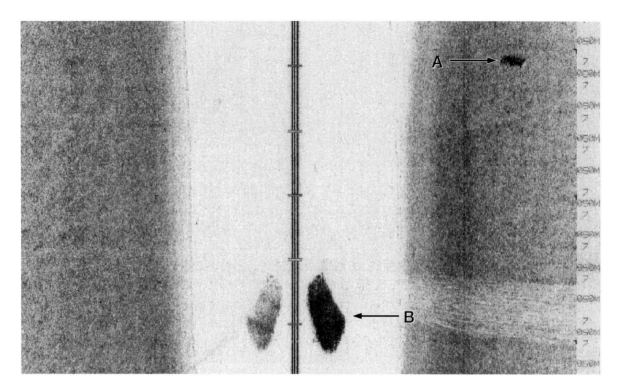

Record 168: This record shows the torn-off wing (A) of a recently lost high wing aircraft. It was first detected on a smooth bottom at an outer range of 200 meters. On repeated runs the image was consistent and repeatable, lending credence to the target's existence. Although the aluminum skin of the wing was intact, it was penetrated by the sound pulse which reflected off the frames inside the wing making the "ribbed" image. The dark target (B) just below and to one side of the transducers is a dense school of fish.

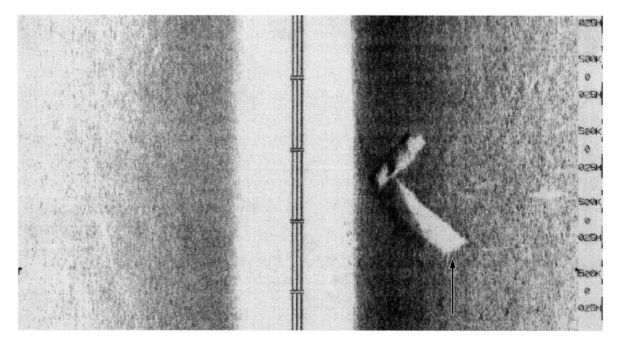

Record 169: This record is of a Grumman AF-2 Guardian, four passenger ASW aircraft. The aircraft was lost over thirty years before the sonar image was generated. Although the tail of the aircraft is now missing, both wings are present with one buried in sediment. The square edge to the shadow (arrow) is more evidence to the shape of the wing than the image of the wing proper.

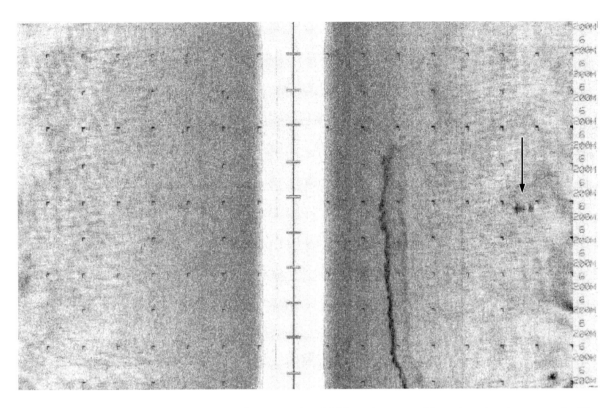

Record 170: This record was generated on a scale of 200 meters per side using slow tow speeds. The target is a downed Cessna 152 aircraft. The dark line in the same channel as the target is a vessel wake.

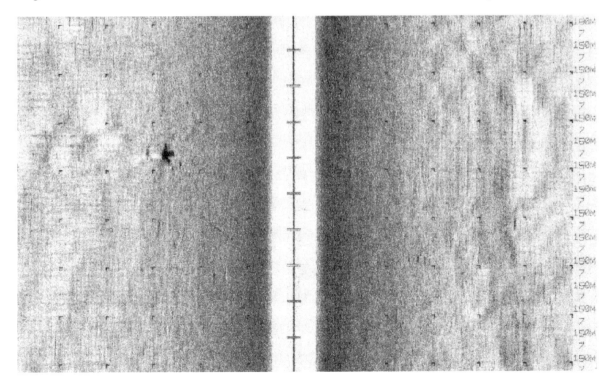

Record 171: This scan of the same downed aircraft in Record 170 was made on a scale of 150 meters. It can be seen that the shorter range of 150 meters provides a more recognizable image and would be preferable to longer ranges on a search for a target of this size.

Appendix A:
Output Pulses per Meter of Forward Motion[*]

SONAR TOW SPEED IN NAUTICAL MILES PER HOUR

	1.0	2.0	3.0	4.0	5.0	6.0	7.0	8.0	9.0	10.0	11.0	12.0
25 M	59.04	29.52	19.68	14.83	11.81	9.84	8.43	7.38	6.56	5.90	5.37	4.92
50 M	29.52	14.76	9.84	7.41	5.90	4.92	4.22	3.69	3.28	2.95	2.68	2.46
75 M	19.68	9.84	6.56	4.94	3.94	3.28	2.81	2.46	2.19	1.97	1.79	1.64
100 M	14.76	7.38	4.92	3.71	2.95	2.46	2.11	1.84	1.64	1.48	1.34	1.23
150 M	9.84	4.92	3.28	2.47	1.97	1.64	1.41	1.23	1.09	.98	.89	.82
200 M	7.38	3.69	7.46	1.85	1.48	1.23	1.05	.92	.82	.74	.67	.61
300 M	4.92	2.46	1.64	1.24	.98	.82	.70	.61	.55	.49	.45	.41
400 M	3.69	1.85	1.23	.93	.74	.61	.53	.46	.41	.37	.34	.31
600 M	2.46	1.23	.82	.62	.49	.41	.35	.31	.27	.25	.22	.20
750 M	1.97	.98	.66	.49	.39	.33	.28	.25	.22	.20	.18	.16

SONAR RANGE PER SIDE

[*]using conventional side scan sonar in marine applications

Appendix B:
Useful Calculations, Conversions and Equalities

1. Calculation of Target Height:

$$H_t = (L_s \times H_f)/R$$

Where H_t is the height of the target, L_s is the length of the shadow cast by the target, H_f is the height of the fish and R is the range from the towfish to the end of the shadow cast by the target.

2. Calculation of actual seabed swath width for a given range setting.

$$SW = 2\,(\sqrt{(R_S^2 - H_f^2)})$$

Where SW is the actual seabed coverage per side, R_s is the range setting of the sonar system and H_f is the height of the fish.

3. Calculation of the layback of the towfish when using short cables. (This calculation does not consider the effect of a catanary on the layback figure. The catanary of the towcable lessens the layback. During operations using short cables, the catanary factor is usually small. For operations using long cables, cable prediction software programs can be used to accurately determine layback.)

$$L = 2\sqrt{(C^2 - D_f^2)}$$

Where L is the layback of the towfish, C is the amount of in-water-cable and D_f is the depth of the towfish.

4. Calculation of pulse width of an active sonar.

$$P_w = C \times P_l$$

Where P_w is the pulse width, C is the speed of sound in water and P_l is the sonar pulse length.

The following conversions and equalities may be useful during search operations but have been rounded to a maximum of three decimal places. Also some equalities may be taken from cartographic principles that are not fixed. Therefore these figures should only be used for close approximations. For high precision survey applications more exact length and area figures may be required. Speed of sound figures depend on other variables and must be considered approximate.

SPEED

TO CONVERT:	MULTIPLY BY:
Knots to feet per minute	101.269
Knots to kilometers per hour	1.852
Knots to statute miles per hour	1.151
Knots to meters per second	0.514
Knots to feet per second	1.688

LENGTH AND AREA

TO CONVERT:	MULTIPLY BY:
Inches to millimeters	25.40
Feet to meters	0.305
Feet to fathoms	0.167
Yards to meters	0.914
Fathoms to meters	1.829
Nautical mile to feet	6076.115
Nautical mile to meters	1852.0
Nautical miles to kilometers	1.852
Cables to feet	720.0
Meters to inches	39.370
Meters to feet	3.281
Meters to yards	1.094
Meters to fathoms	0.547
Kilometers to feet	3280.840
Sq. kilometers to sq. nautical miles	0.292

EQUALITIES:

1 minute of latitude at the equator	=	1842.735 meters
1 minute of latitude at the pole	=	1861.701 meters
1 Kilogram	=	2.205 pounds
1 short ton	=	2000 pounds
1 metric ton	=	2204.623 pounds
1 long ton	=	2240 pounds
Speed of sound in 3.5 % salt water at 15.5 degrees C	=	1507.35 m/s
Speed of sound in fresh water at 15.5 degrees C	=	1472.70 m/s

Appendix C:
Glossary

Absorption: The removal of energy from the sonar beam as it propagates through the water.

Acquisition: The process of detecting and recognizing a seabed anomaly using sonar.

Active Sonar: A sonar system consisting of both a projector and hydrophone, and capable of transmitting and receiving acoustic signals.

Along-track: In a direction parallel to the track of the towfish (transverse).

Altitude: The height of the towfish above the seabed, which is typically measured in feet or meters.

Ambient Noise: Acoustic signals, sensed by the sonar system, emanating from a variety of sources in the underwater environment.

Anechoic: An object or area characterized by an unusually low degree of reverberation; echo-free.

Angle Of Incidence: The angle that a straight line acoustic pulse meeting a surface, makes with a normal to the surface.

Attenuation: The process of weakening or reducing the amplitude of an sonar signal, caused by numerous factors including scattering, beam spreading and absorption.

Backscatter: The deflection of acoustic radiation in a scattering process through an angle greater than 90 degrees.

Bathymetry: The measurement of the depths of oceans, seas or other large bodies of water, typically using narrow swath acoustic systems.

Beam Angle: The amount of rotation needed to bring two opposite sides of a sonar beam, which is diverging from a point (transducers), into coincidence with one another. The beam angle determines the rate of divergence during propagation.

Beam Forming: The process of shaping an acoustic beam through the control of the geometry of the transducer array.

Beam Spreading: The divergence of a sonar beam as a direct function of angle and range.

Beam Width : The distance between two opposite sides of a beam at a specific range from its source.

Blanking : Sonar signal blocking caused by discontinuities in the water resulting in an empty, unprinted space on the sonar record; sometimes, but rarely, caused by local signals or radiation external to the control/display portion of the sonar system.

Boomer: A seismic instrument typically operating in the .5 to 2.5 kHz range, producing a conical beam directed vertically towards the seafloor. Used for profiling geological features beneath the seabed.

Bottom Lock: The method whereby the sonar continuously detects the seabed directly below the towfish and calculates the towfish height. This calculation is important for the slant range correction process in side scan sonar.

Catanary: The curve(s) assumed by a tow cable moving through the water, typically induced by the forces of water drag on the cable.

Cavitation: The rapid formation and collapse of vapor pockets in water, most often caused by a significant and rapid drop in pressure. Some cavitation bubbles do not redissolve rapidly and are a major cause of quenching of the sonar signal.

Channel: One of two or three signals in a multi-signal sonar system; the area on the display or sonar record where data from this signal is shown.

Compression : A single axis reduction in size of a sonar image due to speed or slant range distortions.

Correction: The process of removing errors caused by speed, slant range or other sonar distortions.

Coverage: An area described by the seabed swath width of a side scan sonar and the distance traveled by the survey vessel on its track; also pertains to the repeat surveying of an area i.e. one pass equals 100% coverage of an area and 2 passes over the same area equals 200% coverage.

Cross-track : The direction 90 degrees to the path of the vessel or towfish; the range dimension.

Deadweight Depressor: A heavy, inert weight used to increase towfish depth when attached to the tow cable.

Decibel: A unit used to express the intensity of a sound wave, equal to 20 times the common logarithm of the ratio of the pressure produced by the sound wave to a reference pressure.

Depressor: An attachment to a sonar tow cable that assists in increasing the depth of the towed body; commonly of two types: deadweight and hydrodynamic.

Detectability : The size, shape and makeup of a seabed anomaly as related to a sonar's ability to discern its existence.

Differentiation: The process of using separate but identical navigational instruments where one is fixed at a known location and provides, via radio link, a second mobile instrument with offset calculations. This process is used to increase the accuracy of certain navigational instruments that may be affected by diurnal or atmospheric variations.

Discontinuity : A change in the make up of a body-of water that causes a change in the speed, and/or direction of sound propagation, of an incident sonar pulse.

Drag: The hydrodynamic forces exerted on the components of a tow assembly that tend to reduce its forward motion.

Event : A mark or notation put on a sonar record, or embedded in stored data, representing the moment of a navigational fix or other critical occurrence during a survey.

First Bottom Return: The component of a side scan sonar record representing the shortest acoustic path between the towfish and the seabed directly below the towfish.

First Surface Return: The component of a side scan sonar record representing the shortest acoustic path between the towfish and the surface directly above the towfish.

Fish Height: The distance between the towfish and the seabed, usually measured in feet or meters.

Footprint: The area of seabed affected by the increase in the level of sound from an outgoing sonar pulse during, or after, a specific period of time.

Frequency: The number of cycles or completed alternations per unit time of a sound wave, most often measured in Hertz. Frequencies commonly used in conventional side scan sonar range from 25 to 450 kHz.

Fully-corrected: A speed and slant range corrected sonar record which accurately depicts the seafloor in a 1:1, two dimensional image.

Gain: A measure of the increase in signal amplitude produced by an amplifier, most often through time-varied-gain circuitry, or printer controls.

GPS : (Global Positioning System) A satellite based navigation system providing accuracies usable for side scan sonar surveys on a worldwide basis.

Grazing Angle: The angle at which the side scan sonar pulse strikes, and propagates across, the seafloor.

Heave: The rise and fall of a surface vessel or towfish in a rhythmic movement; the disjointed, jagged images on a sonar record produced by towfish heave.

Horizontal Beam Width: The angle of the transmitted (and/or received) sonar beam in the along-track (transverse) dimension, often between .5 and 2.5 degrees for side scan sonar.

Hydrodynamic Depressor: A tow assembly depressor designed with louvers or vanes oriented in such a way as to increase negative lift when exposed to an increased water flow.

Hydrophone: a sonar receiver functioning by transforming underwater sound signals (pressure waves) into electrical signals.

Hz: A unit of frequency equal to 1 cycle per second named after the physicist, H.R. Hertz (1857-1894).

Insonify (Ensonify, Br.): To expose an area, or portion of seabed, to sonar energy.

Instability : The behavior of a towfish experiencing the motion of heave, pitch, roll or yaw; the erratic motion of any part of a towing assembly usually resulting from fluctuating drag forces.

Interference: The display of erroneous signals from acoustic or electrical sources that conflict with the display of the primary side scan sonar data.

Interferometric Sonar: A system based on the process by which two or more sonar waves of the same frequency combine to reinforce or cancel each other, the amplitude of the resulting wave being equal to the sum of the amplitude of the combining waves. Because the angle of interference can be determined, these sonar systems provide bathymetric information simultaneously with side scan data.

In-water-cable: The amount of tow cable exposed to part or all of the water column and potentially affected by drag forces.

Kiting: A rhythmic, lateral movement experienced by tow bodies on long cables and in deep water; most often induced by poor hydrodynamics of a depressor.

Lane: A course or track, down the center of which, a survey vessel travels during a survey. A lane is delineated on either side by half the distance between the current and adjacent tracks.

Lane Spacing: The distance between successive vessel tracks in a multi-lane survey.

Lateral: To the side, or in the cross-track dimension, of the towfish path.

Layback: The horizontal distance between the survey vessel, or the navigation antenna, and the towfish.

Loran: A navigation system having accuracies of 100's of meters, based on the time displacement between signals from two or more fixed shore based antennas.

Mapping: Creating sonar records that accurately represent 1:1 scaler plan views of large sections of seabed; also creating high resolution images of complex underwater targets.

Mensuration: Scaling of the physical dimensions and volume of an object from a sonar record.

Microbars: A unit of pressure equal to one millionth of a bar, commonly used to indicate acoustic signal strength.

Mosaic: An assembly of sonar records matched in such a way as to show an accurate, continuous, 2 dimensional representation of an area of seabed.

Multipath: Sonar signals arriving at a target, or the towfish, from a single source but along different paths.

Noise: Extraneous signals detected by a sonar that affect the system's efficiency to display, and the operator's efficiency to interpret, the signals of interest.

Out-of-range: Target echoes displayed by the sonar, resulting from hard reflectors that are beyond the system range setting.

Overlap: The area of seabed that is covered a second or more times, referred to as a percentage of swath width.

Over-the-ground: A measurement of speed of the survey vessel or towfish as a true speed over the seabed, independent of movement in relation to wind or water.

Pass: A single procession near a seabed anomaly during a sonar survey.

Passive Sonar: A sonar system having only a hydrophone and capable of receiving signals but not transmitting them.

Path-tracking: The ability of a towed body to accurately follow the path along which it is towed by a surface vessel.

Ping: A single output pulse of a sonar system; also the returns from a single output resulting in the lateral display of one individual line of side scan data.

Pitch: An instability in the towfish expressed by the alternate rise and fall of the nose and tail about a horizontal axis.

Plan View: A to-scale side scan sonar display constructed to represent the top view of a section of seabed.

Post-processing: Sonar data processing after real time data generation and storage.

Pre-plots: Specific points including tracklines and navigation fix points, the positions of which are determined prior to the commencement of a survey.

Profiler: An instrument that records a vertical section, or simple outline, of the seafloor along a survey line.

Projector: A sonar transducer that translates an electrical signal into pressure waves (sound signals) and transmits them through the water.

Propagate: The movement of sound waves through the water; also transmit.

Pulse: A short burst of sonar, typically measured as a function of time, distance or power.

Pulse Length: The length of time that an active sonar is transmitting one pulse, typically expressed in milliseconds.

Pulse Width: The thickness of the insonified water, in the range dimension, at a given point in time, expressed in meters and determined by multiplying the pulse length by the speed of sound through the water.

Quench: The loss of a sonar signal, most often due to water borne discontinuities and resulting in blank sonar display areas.

Q-Route: A route of safe passage through a mined waterway.

Range: A sonar setting which represents a distance, usually measured in meters, that is the maximum distance from the towfish that the sonar will display (the range setting on the sonar also determines the time between outgoing sonar pulses); also synonymous with the cross-track dimension.

Range Resolution: The ability of the sonar to image, separately and distinctly, objects that lay in a line 90 degrees to the towfish heading.

Range Data Compression: Sonar image compression resulting from the geometry of slant range side scan sonar displays.

Range Overlap: The area of seabed, lateral to the towfish track, reinsonified on successive tracks during a survey; equal to the swath width less the lane spacing, usually expressed in meters.

Ray Bending: Changes in the speed and direction of a sonar beam in the water.

Recognition: The acknowledgment by the sonar operator of the existence of a target or anomaly as displayed in the sonar data.

Refraction: The change of direction of a sound beam when passing obliquely from one medium into another, where its wave velocity is different.

Reverberation: The echoing of a sonar signal from a target or targets.

Roll: The rhythmic movement of a ship or towbody about its longitudinal axis.

Rub-Test: The process of manually creating friction on a transducer face in order to test system electrical continuity.

Sea clutter: The images created in a sonar display by acoustic returns from a rough sea surface.

Scale Marks: Equidistant, regular marks on a sonar display used to assist in the mensuration of targets and anomalies and to provide information on the range displacement of targets from the towfish path.

Scattering: The diffusion of the sonar signal in many directions through refraction, diffraction and reflection, primarily due to the material properties of the insonified areas.

Shadow: A light area on a normal sonar record that is less insonified than the surrounding region; most often caused by signal blocking from an acoustically opaque object on or above the seafloor.

Slant Range: The straight-path time of arrival of a sonar signal along the hypotenuse of a triangle described by the towfish, the seafloor directly below it, and the seabed point of interest.

Slant Range Correction: A computerized repositioning of sonar data on the display to counteract range data compression.

Slip Ring: An electromechanical component, most often used on a winch, that allows full electrical continuity of a sonar cable during winch drum operation.

Sonar Geometry: The spatial relationship between the sonar transducers and their environment including the seafloor, targets and the sea surface.

Sonograph: A hard copy display of sonar data generated either in real time or from recorded data.

Specular Reflector: An object, to which incident sonar beams are largely normal to its surface, making it a strong reflector from a variety of angles. Objects in this category include cylindrical objects such as pipes, orthogonally positioned plates and spheres.

Speed Correction: The proportional matching of sonar chart length with the OTG speed of the survey vessel.

Swath Width: The lateral coverage of side scan sonar on the seabed.

Termination: The junction of either end of a towcable where it is fitted with a single or multi-pin connector.

Thermocline: A layer of water where the vertical temperature gradient is greater than that in the water above it or in the water below it.

Time Varied Gain: (TVG) A process where amplifier gain is changed based on time and matched with the returning signals between outgoing pulses of a side scan sonar.

Transducer : The electromechanical component of a sonar system that is mounted underwater and converts electrical energy to sound energy and vice versa.

Transverse Resolution: The ability of the sonar to image, as separate and distinct, objects that lay in a line parallel with the towfish track.

Trigger Pulse: The signal provided to sonar transducer firing circuitry to initiate the outgoing pulse; also two parallel lines on the center of a sonar record that represent the position of the fish in relation to the sonar image.

Vertical Beam Width: The angle of the transmitted (and/or received) side scan sonar pulse in the vertical dimension, typically between 40 and 70 degrees.

Water Column: The vertical section of water from the surface to the bottom in which a sonar may be towed; also the center section of an uncorrected sonar record.

Wavelength: The distance, measured in the direction of propagation, between two successive points in a wave that are characterized by the same phase of oscillation.

Yaw: An instability characterized by the side to side movement of a ship or towed body about its vertical axis.

Z-Kinking: The failure of cable conductors (characterized by the "z" shape of the damaged portion) resulting from apparent movement between the core and the jacket. At underwater cable terminations, this can occur with conductor extrusion under pressure (pistoning) from a flexible jacket into the dead end of a connector body.

Bibliography

Albers, V. M.; *"Underwater Acoustics Handbook"*, Pennsylvania State University Press, PA, 1961.

Browning, David, Scheifele, Peter, Mellen, Robert; **"Attenuation of Low Frequency Sound in Ocean Surface Ducts: Implications for Surface Loss Values"**; *Proceedings-Oceans '88 Technical Conference, Marine Technology Society*; 1988.

Clay, C. S. and Medwin, H.; *"Acoustical Oceanography: Principles and Applications"*, Wiley-Interscience, New York, 1977.

Cloet, Roger; **"Implications of Using a Wide Swathe Sounding System"**; *Proceedings-Oceans '88 Technical Conference, Marine Technology Society*; 1988.

Cole, F. W. *A Familiarization With Lateral or Side-Scanning Sonars*: Hudson Laboratories, Dobbs Ferry, NY, 1968.

Driscoll, Alan; *Handbook of Oceanographic Winch, Wire and Cable Technology* Ocean Technical Systems, N. Kingston, RI, 1989.

Dyka, Andrzej; *"Resolution Improvement Filter for Sonar Returns"*, Technical University of Gdansk, Poland, 1987.

Flemming, B.W.; **"Side Scan Sonar; A Practical Guide"**; *The International Hydrographic Review*, vol 53, no 1; 1976.

Flemming, B.W.; *"Causes and Effects of Sonograph Distortion and Some Graphical Methods for Their Manual Correction"*; National Research Institute for Oceanology, Department of Geology, University of Cape Town, South Africa.

Flemming, B. W.; *Recent Developments in Side Scan Sonar Techniques.* Central Acoustics Laboratory, University of Cape Town, South Africa. 1982.

Fish, J. P.; **State of The Art Sonar Images.** *Skin Diver*, Los Angeles, CA: Peterson Publishing Co., 1983.

Fish, J.P., M. Klein; **"Reconnaissance Hydrographic Surveys for Establishment of Navigation Data in Third World Countries"**; *Papers of the 2nd International Hydrographic Technical Conference*; Plymouth, England; 1984.

Fish, J.P.; **"Dive Site Inspection Using Side Scan Sonar"**; *Proceedings of the 7th Meeting of the United States-Japan Cooperative Program in Natural Resources, Panel on Diving Physiology and Technology*; Japan Marine Science and Technology Center, Tokyo Japan; 1983.

Fish, J.P.; *Unfinished Voyages: A chronology of shipwrecks in the Northeastern United States, 1606-1956*, Lower Cape Publishing, Orleans, MA; 1989.

Fish, J.P. and Carr, H.A. *Seabed Target Detectability Using Wide Area, High Speed Sonar Search Methods - Technical Note Number 12*. AUSS, Cataumet, MA; 1987.

Fish, J.P. and H.A. Carr **"Integrated Remote Sensing of Dive Sites"**; *Proceedings - Oceans '88 Technical Conference, Marine Technology Society*; 1988.

Fish, J.P. and H.A. Carr; **"High Resolution Acoustic and Optical Imaging of Discrete Underwater Targets"**; *Proceedings of Subsea Salvage 90, Hollywood, FL*; May 1990.

Fish, M. P., Kelsey, A. S., Mowbray W. H.;**"Studies on the Production of Underwater Sound by North Atlantic Coastal Fishes"**; *J. Marine Research*,11, 1952.

Forbes, S. T. and Nadden, O.; *"Manual of Methods for Fisheries Resource Survey and Appraisal; Part 2. The Use of Acoustic Instruments for Fish Detection and Abundance Estimation"*, FAO, Rome Italy, 1972.

Guieysse, L. and Sabathe, P.; *"Acoustique Sous-Marine"*, Dunod, Paris, 1964.

Hersey, J. B.; **'A Chronicle of Man's Use of Ocean Acoustics"**; *Oceanus*, Vol. 20, No. 2; 1977.

Horton, J. W.; *"Fundamentals of Sonar"*, U.S. Naval Inst., Annapolis, 1957.

Hubbard, Robert; **"Hydrodynamic Design of a Sonar Towfish for Minimal Image-Smearing in Heavy Seas"**; *Proceedings-Oceans '89; Technical Conference, Marine Technology Society*; 1989.

Hunt, F. V.; *"Electroacoustics: The Analysis of Transduction and its Historical Background"*, Harvard University Press, 1954.

Johannesson, K. A. and Mitson, R. B.; *"Fisheries Acoustics. A Practical Manual for Aquatic Biomass Estimation"*, FAO, Rome Italy, 1983.

Kinsler, L. E. and Frey, A. R.; *"Fundamentals of Acoustics"*, Wiley, New York, 1962.

Lesnikowski, Nicholas; **"Deep Towed Interferometric Swath Bathymetry"**; *Proceedings-Oceans '89; Technical Conference, Marine Technology Society*; 1989.

Morse, P. M.; *"Vibration and Sound"*, McGraw-Hill, New York, 1948.

Newton, Fred Jr.; *"The Consummate Search/Survey System - New Developments in Sonar Image Processing"*; Triton Technology, Inc., Watsonville, CA; 1989.

Nicholson, John, Jaffe, Jules; **"Side Scan Sonar Acoustic Variability"**; *Proceedings-Oceans '88 Technical Conference, Marine Technology Society*; 1988.

Nunnallee, E. P., *"An Alternative Method of Thresholding during Echo Integration Data Collection"*, National Marine Fisheries Service, Washington, 1987.

Porter, Robert; **"Acoustic Probing of Ocean Dynamics"**; *Oceanus*, Vol. 20, No. 2; 1977.

Robinson, Larry, Oyvind Bjorkheim; **"Interferometry; An Alternate Method in Sonar Mapping"**; *Proceedings-Oceans '89; Technical Conference, Marine Technology Society*; 1989.

Ryan, Paul; "**A Readers Guide to Underwater Sound**"; *Oceanus*, Vol. 20, No. 2; 1977.

Ryther, J.H. Jr., D.B. Harris and J.P. Fish **"Putting ROVs To Work Investigating Shipwrecks"**; *Sea Technology*; 1990.

Smircina, K.L. and J.P. Fish; *"Remotely Operated Vehicle (ROV) Reliability Study, Phase I"*; Limited Distribution, Prepared for Office of Naval Research, Washington D.C.; 1987.

Spindel, Robert; "**Acoustic Navigation**"; *Oceanus*, Vol. 20, No. 2; 1977.

Stanta, T. K.; **"Effects of Transducer Motion on Echo Integration Techniques"**, *Journal of the Acoustic Society of America*, 1982.

Stefanon, Antonio; **"Marine Sedimentology Through Modern Acoustical Methods: 1. Side Scan Sonar"**; Vol. III N.1, Gennaio 1985.

Tucker, D.G. and Gazey, B.K.; *"Applied Underwater Acoustics"*, Pergamon, 1966.

Urick, R.J.; *"Principles of Underwater Sound"*; McGraw-Hill, New York, 1975.

Vine, Allyn; "**Sight Through Sound**"; *Oceanus*, Vol. 20, No. 2; 1977.

Index

TEXTUAL SUPPLEMENTS

This text is designed to be supported by illustrated supplements to be published on a variety of sonar subjects including: deep side scan sonar operations, bathymetric side scan systems, towed body positioning, surface vessel positioning, sonar data storage and sonar image processing.

These supplements are planned for a biannual availability and are designed to keep the reader abreast of technological developments in the field of side scan sonar.

If the reader would like to be advised when each of these supplements becomes available, this form should be completed and returned to:

AUSS LTD.
Box 768-A
Cataumet, MA USA 02534.

Yes, Please inform me by mail when illustrated supplements to *Sound Underwater Images* become available.

Name: _____

Title: _____

Organization: _____

Street or
Post Box: _____

Town: _____

Country: _____

Postal Code: _____

Tel: _____

Fax: _____